THINKING ABOUT BIOLOGY

AN INTRODUCTORY LABORATORY MANUAL
SECOND EDITION

MIMI BRES • ARNOLD WEISSHAAR
PRINCE GEORGE'S COMMUNITY COLLEGE

PEARSON

Prentice
Hall

Upper Saddle River, NJ 07458

Editor-in-Chief, Science: John Challice
Executive Editor: Gary Carlson
Project Manager: Crissy Dudonis
Vice President of Production & Manufacturing: David W. Riccardi
Executive Managing Editor: Kathleen Schiaparelli
Assistant Managing Editor: Becca Richter
Production Editor: Rhonda Aversa
Supplement Cover Manager: Paul Gourhan
Manufacturing Buyer: Ilene Kahn
Photo Image Credit: Art Wolfe/Getty Images, Inc.

© 2005 Pearson Education, Inc.
Pearson Prentice Hall
Pearson Education, Inc.
Upper Saddle River, NJ 07458

Printed in the United States of America

10 9 8 7 6 5 4 3 2 1

ISBN 0-13-145820-5

Pearson Education Ltd., *London*
Pearson Education Australia Pty. Ltd., *Sydney*
Pearson Education Singapore, Pte. Ltd.
Pearson Education North Asia Ltd., *Hong Kong*
Pearson Education Canada, Inc., *Toronto*
Pearson Educación de Mexico, S.A. de C.V.
Pearson Education—Japan, *Tokyo*
Pearson Education Malaysia, Pte. Ltd.

Contents

Preface

Thinking About Biology is designed for a one-semester, nonmajors, general biology course with a human focus. The topics and exercises are general enough to be compatible with most introductory level human and general biology texts currently in use. The activities demonstrate that basic biological concepts can be applied to a wide variety of plants, animals, and microorganisms.

This book is unique not because of the specific topics covered, but because of the approach to these topics. These laboratory exercises are planned to help your students

- gain practical experience that will help them understand lecture concepts,
- acquire the basic knowledge needed to make informed decisions about biological questions that arise in everyday life,
- develop the problem-solving skills that will lead to success in school and in a competitive job market,
- learn to work effectively and productively as a member of a team.

The basis of scientific work is asking questions and answering them by observations or experiments. Thus, we strongly feel that the most important goal of an introductory course in the life sciences should not be to simply outline facts about animals, molecules, or new techniques. Rather, students should come away with some understanding of the processes of investigation that are basic to science, and how scientists work to solve problems.

We hope that working through this laboratory manual will be an exciting experience for both students and instructors, and one that will leave students better prepared to meet the demands of our increasingly scientific society.

SPECIAL FEATURES FOR YOU, THE INSTRUCTOR

Three-Pronged Approach to Laboratory Learning

- **Elicit Interest.** Students perform better when they have a clear understanding of the reason and objectives behind each exercise. Students are often more willing to proceed if they know why they are asked to do each activity and what objectives they are expected to accomplish during the laboratory period.
- **Clear Directions.** Students need clearly and concisely written directions of how to proceed. This is the first and last science class for many students and we can make very few assumptions about lab skills they might bring into the class.
- **Establish Relevance.** A well-designed lab needs to help students understand the meaning and importance of the findings they just gathered, and to establish relevance between laboratory and lecture topics and with everyday life.

Clarity

- All content and problem solving goals are accomplished by using a **simple, non-threatening approach** to the lessons. Even students with little science background are able to **get involved** and master the material.

Instructional Objectives

- A clear set of **Instructional Objectives** begins each exercise. This helps students identify the most important concepts that will be emphasized in each activity.

Focus on Thinking and Problem Solving

- One of the most important features of this laboratory manual is the **infusion of thinking and problem-solving skills with the content areas**. Research has shown that it is easiest to master thinking and problem-solving skills when thinking and content are learned together. **Thinking About Biology** is based on this approach. Activities and questions within an exercise build on previously learned information and encourage students to transfer information from one section of the course to another.

Team Building Opportunities

- Most laboratory activities emphasize a **team approach**. Group work is encouraged and often required. In the real world, we are expected to interact with others to solve problems and complete projects. This approach provides opportunities for students to work together, share ideas, and function effectively in groups to accomplish tasks.

Real-life Connections

- Questions throughout the activities and those incorporated into the **Self Tests** are designed to connect the laboratory activities to timely subjects and real-life experiences. The exercises are intended to stimulate interest in topics that can help them make decisions regarding their own health and nutrition, understand current topics in the news, and their role in the environment.

Many Opportunities to Practice Skills and Reinforce Concepts

- Activities are **reinforced with relevant questions**, many more than can be found in any competing editions. It is not only the number of questions, but also the **type** of questions that are important. Questions are designed to provide **continuous feedback** and **elicit targeted thinking**. This keeps the learner "on track" and fosters development of problem-solving skills.
- Through **questions interwoven** into each activity, **Comprehension Checks**, and **Self Tests**, recall of important concepts is reinforced. Students also have the opportunity to transfer content knowledge gained to new situations.

These Lessons Work!

- These laboratory activities have been **fully tested** with more than 15,000 students on this and other campuses around the United States.

■ Experience has shown that the simplified structure and easy-to-understand language of this manual works well with a diverse student population.

Requires Minimal Supplies and Equipment

■ These exercises are cost-effective and affordable even for schools with budget constraints. **Most schools will already have all the equipment needed to perform the lab activities.**

Designed for Simplicity

■ All activities are **easy to set up and maintain**. Our program is large and multisectional. Despite large numbers of instructors and students, this laboratory curriculum is incorporated smoothly in all sections. The simplicity of the laboratory approach has allowed new personnel to be **"up and running" with minimal preparation time.**

■ Since the exercises require minimal equipment, many activities throughout the manual are **suitable for use in on-line sections.**

Adaptable to Any Curriculum

■ Each exercise is designed for use in a three-hour laboratory period, but can easily be adapted to accommodate two-hour or 90-minute sessions. To provide **maximum flexibility** for instructors, each exercise is **divided into several activities**. Activities can be deleted or presented as demonstrations without diminishing the value of the remaining components.

■ Laboratory exercises can be presented **in sequence or rearranged** to suit your needs. The manual includes 20 exercises, allowing you the latitude to tailor the laboratory curriculum to your course and the equipment available at your institution.

Support Materials - Instructor's Guide.

The Instructor's Guide provides you with

■ complete preparation and setup instructions for each activity,
■ a complete Answer Key for all activity questions,
■ suggested schedule for split labs,
■ helpful hints to facilitate smoothly running laboratories,
■ master charts for recording results of classroom experiments.

SPECIAL FEATURES FOR YOU, THE STUDENT

Informal Style

■ The text is written in an active voice with easy-to-understand language. Key terms and important definitions are highlighted with bold print for easy recognition. Large spaces are provided throughout the manual for you to record and fully explain your answers.

Active Learning Experience

- Every exercise gives you an opportunity to be an active participant in the scientific method. You will form hypotheses, set up experiments, collect data, record your data in graphs and charts, and draw conclusions from your experimental results.

Self-guided Approach

- The laboratory procedures are clearly outlined to allow you to progress through each activity independently.

Tools for Success

The following components of each exercise will help you to succeed:

- **Instructional Objectives.** The objectives are listed first in each exercise so that you will be able to focus your attention on the main concepts of each activity.
- **Content Focus.** Each exercise includes a brief discussion of the background information that you will need to understand the subject of the exercise and prepare you to complete the activities that follow.
- **Notes and Cautions.** Note boxes provide helpful hints for solving problems and accomplishing laboratory tasks. Pay special attention to caution boxes that provide important safety information.
- **Comprehension Checks.** Stop and complete the Comprehension Check questions to get immediate feedback on your understanding of the basic principles covered in each activity. These questions also provide a chance for you to apply what you have learned to situations outside the classroom.
- **Check-off Boxes.** These boxes allow your instructor to check your progress before you move on to a new activity.
- **Self Tests.** Answer these questions after completing the laboratory exercise. Self Test questions allow you to assess your comprehension and apply your knowledge. Answers to the Self Test questions can be found in Appendix I at the back of the book.

ACKNOWLEDGMENTS

Any project of this size and scope cannot be completed without support and cooperation from many people. We would like to express our deep appreciation for their contributions.

Thanks to

Michael Garvey, Federal Bureau of Investigation, Washington, D.C., for providing your time and technical expertise in the area of forensic molecular genetics.

Kenneth Thomulka, Assistant Professor of Biology, Philadelphia College of Pharmacy and Science, for allowing us to adapt an imaginative and unique application of biotechnology to the education arena.

The Biology 101 faculty at Prince George's Community College. Your many helpful suggestions have contributed substantially to the effectiveness of these laboratory exercises.

Sandra Dempsey, Biology Laboratory Manager for facilitating the testing of new laboratory activities.

Charlotte Weisshaar, for proofreading and technical support in the development of this manuscript.

Robert Ewing and Margaret Ryan, for the excellent artwork and photographs you contributed to this manual.

Marie Robinson, Office Goddess, Science Department, Prince George's Community College, for doing the many little things that are necessary to put a publication together—and doing it with humor!

The Science Department office staff, Barbara Blum and Ai Nguyen, for their technical support.

The Biology 101 students for classroom-testing our laboratory exercises and offering suggestions for improvements.

Crissy Dudonis, Project Manager for Biology, Prentice Hall, for lending your skill and support to this project. Your efficient and effective management played a significant role in making this second edition possible.

Numerous reviewers listed below graciously gave their time to comment on the manuscript and suggested many useful changes. For their help, many thanks.

REVIEWERS

John S. Campbell, *Northwest College*

Renee E. Carleton DVM, *Berry College*

Dwight O. Kamback, *Lake City Community College*

Will Kleinelp, *Middlesex County College*

H. Roberta Koepfer, *Queens College, CUNY*

Monica McGee, *University of North Carolina- Wilmington*

William Simcik, *Tomball College*

Jason Yoder, *Itasca Community College*

EXERCISE 1

Introduction to the Scientific Method

Objectives

After completing this exercise, you should be able to:

- use the scientific method to solve problems
- organize information to facilitate analysis of your data
- draw graphs that present data clearly and accurately
- interpret data in tables, charts, and graphs
- draw conclusions that are supported by experimental data
- analyze data using common statistical measures
- apply your knowledge of the scientific method to real-life situations.

CONTENT FOCUS

What is science? What do scientists *do* all day? These are not easy questions for most of you to answer. The widespread idea of a nerd in a white lab coat does not apply to most scientists. So, what are scientists really like? They all have the "**three Cs**" in common.

Just as you are, scientists are **curious** about the world around them. They ask questions about everything. Can my diet cause heart disease? Why does the river look brown instead of blue? How can squirrels remember where they bury their nuts? Why do some cars get better mileage than others? Science is a method for answering these and many other questions.

Scientists don't accept things without **collecting information**. All the facts relating to a problem or question have to be carefully explored and checked for accuracy.

Scientists are **comfortable with new concepts**. If a better explanation can be found, scientists are not afraid to give up old ideas for new ones.

1

To make the three Cs happen, scientists have developed a series of steps in investigation called **the scientific method**. Through trial and error, the scientific method has proven to be an efficient and effective way of attacking a problem. You have probably used some version of the scientific method many times in your life—without being aware of the steps you were following.

ACTIVITY 1. FORMING HYPOTHESES: DARING TO BE WRONG

Preparation

There are several different ways that a problem can come to your attention. Someone may **assign** you the problem (this happens often in a school or a work situation), the problem may **thrust itself** upon you (your car won't start), or you may discover the problem by simply being **curious** about something you have seen.

Let's begin with a simple situation that you might face any day.

> ### The Problem
> **You drive to school and park in your usual spot. As you walk across the campus after your morning biology class, you discover that you can't find your car keys. You have a problem!**

An easy way to attack the problem is to make an **educated guess** about the possible solution to the problem. It is an "educated" guess because you use all the background information that is available when making your guess.

In scientific terms, an educated guess is called a **hypotheses**.

1. On the next page, in Table 1-1, you will find some hypotheses that might shed some light on this problem.

2. **Complete the table** by adding some hypotheses of your own.

Check your hypotheses with your instructor before you continue.

TABLE 1-1
KEY LOSS: POSSIBLE HYPOTHESES
A: I didn't bring the keys with me today.
B: The keys are in my book bag.
C:
D:
E:

ACTIVITY 2 FORMING HYPOTHESES TO SOLVE PROBLEMS

The Problem

You were absent from chemistry class the day your professor gave out the instructions for making an important solution needed for your laboratory experiment. No problem. Your roommate was in class and had copied down the formula for you. You rush off to chemistry lab and prepare the solution, but when you use it in your experiment, it doesn't perform as expected.

In **Table 1-2, list three hypotheses** about why the formula didn't work. Don't forget— the hypotheses must be **testable**!

T A B L E 1 - 2 **CHEMISTRY EXPERIMENT: HYPOTHESES**
A:
B:
C:

 Check your hypotheses with your instructor before you continue.

Some hypotheses can be tested by observation only, but more often, you will need a **combination of observation and experimentation** to be sure about the accuracy of your results. To understand how scientists work, you must follow the steps of the scientific method as they are used in actual **experiments. In Activities 3 and 4**, you will **see how scientific method skills** are used to **set up experiments** and analyze the **information (data)** that is collected.

ACTIVITY 3   TESTING HYPOTHESES

The Problem
Investigate the effects of fertilizer on plant growth.

STEP 1:

You form a hypotheses about what you think will happen.

Hypotheses: Adding fertilizer will make plants grow taller.

STEP 2:

You design an experiment that compares the growth (in height) of plants that **receive fertilizer** with those grown **without fertilizer. Your design might be similar to the following:**

Begin with **20 plants** (same size, same type), planted in the same-size pots, with the same amount and type of soil, placed on a windowsill with the same exposure to light. All these factors will be held constant.

■ An experiment is designed to isolate the factor you are interested in testing. All **other conditions must be held constant**. In this way, you are sure that your **observed results** were caused by the only factor that was varied.

■ Since you are investigating the **effect of fertilizer**, you will want to hold all **other factors constant** (plant type, plant size, pot size, amount of water, amount of light, etc.) to avoid confusion. This way you can be **sure** that any differences in height are **due to the presence of fertilizer** and not some other factor.

You decide to measure the growth of your plants (height in centimeters) **once a week for a month**. You will keep detailed **records** of your **observations**.

It is helpful to plan an experiment with a **group** of plants (or animals). There are **two good reasons** to use groups:

■ If **unexpected factors** (such as disease) affect one or two experimental subjects, it will not ruin the experiment.

■ **Natural genetic variability** will cause some plants to grow taller than others (just as some people grow taller than others). You can separate this effect from that of the fertilizer by measuring the height in a **group** of plants for each treatment (fertilizer and no fertilizer).

You decide that **10 plants will receive identical measured amounts of fertilizer** each week. These are the **experimental** plants. They are receiving the treatment (fertilizer) that will help you test your original hypotheses (does fertilizer make plants grow taller?).

Ten plants will receive no fertilizer. These are the **control** plants. They do not receive the experimental treatment. You will use these for **comparison with the experimental group** to help you interpret your results, and to show that any observed differences in height between the two groups are due to the **only difference between them**—application of fertilizer.

In the above example, there are **10 replications of the experimental treatment and 10 replications of the control treatment.**

✓ Comprehension Check

1. Why is it necessary to divide the plants into two groups (a control group and an experimental group)?

2. Why is it important to keep conditions exactly the same in the control and experimental group, **except** for the application of fertilizer?

3. Why is it better to do the plant/fertilizer experiment with **10** plants in each group instead of just one or two? Give **two** reasons. **Use your own words**.

    a.

    b.

Check your answers with your instructor before you continue.

ACTIVITY 4   INTERPRETING DATA

Preparation

The month is up. You are ready to draw conclusions from your **data** (the information you have recorded). You will be thinking about what your results mean and whether your hypotheses is **supported**.

The information you collected during your experiment is presented in **Tables 1-3 and 1-4**.

✓ Comprehension Check

1. Were there differences in growth between the control and experimental plants?

   If so, which group grew taller? How do you know?

2. Do the results support the original hypotheses? **Explain** your answer.

3. Why is it more accurate to compare the **average** height gain of the control and experimental groups (instead of comparing individual plants)?

Check your answers with your instructor before you continue.

T A B L E   1 - 3

HEIGHT GAIN (cm) OVER FOUR WEEKS—CONTROL PLANTS

Plant Number	Initial Height (cm)	Week 1	Week 2	Week 3	Week 4	Total Height Gain	Avg Height Gain
1	10.0	1.6	2.0	3.0	2.5	9.1	2.3
2	11.5	2.2	1.5	1.5	2.0	7.2	1.8
3	9.6	1.5	2.3	2.6	2.0	8.4	2.1
4	9.2	2.0	3.0	2.8	1.5	9.3	2.3
5	10.2	2.3	1.2	1.6	2.0	7.1	1.8
6	11.0	3.2	1.7	2.0	3.2	10.1	2.5
7	10.0	2.6	3.0	3.0	1.4	10.0	2.5
8	9.7	4.0	2.6	4.0	2.3	12.9	3.2
9	10.4	DIED	—	—	—	—	—
10	10.4	2.3	2.3	2.7	2.0	10.0	2.5
TOTAL						**84.1**	**21.0**

T A B L E   1 - 4

HEIGHT GAIN (cm) OVER FOUR WEEKS—EXPERIMENTAL PLANTS

Plant Number	Initial Height (cm)	Week 1	Week 2	Week 3	Week 4	Total Height Gain	Avg Height Gain
1	9.6	4.2	5.0	3.0	4.7	16.9	4.2
2	9.8	6.0	4.0	5.5	5.0	20.5	5.1
3	10.3	5.3	5.5	3.6	4.2	18.6	4.7
4	11.0	2.1	3.2	6.2	3.8	15.3	3.8
5	10.1	3.4	4.0	4.4	4.0	15.8	4.0
6	9.2	4.7	3.1	3.1	4.0	14.9	3.7
7	9.5	4.2	5.2	3.9	3.6	16.9	4.2
8	10.0	3.3	6.0	5.6	4.2	19.1	4.8
9	9.7	5.8	6.1	6.5	5.0	23.4	5.9
10	10.4	5.1	3.4	5.8	5.3	19.6	4.9
TOTAL						**181.0**	**45.3**

ACTIVITY 5   FORMING HYPOTHESES FOR AN EXPERIMENT

Preparation

Now that you have had some experience using the steps of the scientific method, you will develop a hypotheses and test it in a simple experiment.

Table 1-6 contains some examples of physical traits and physiological factors that may or may not be related. You can choose to investigate **any two factors from Table 1-6.**

1. **Working alone,** form a **hypotheses** to test in your experiment.

2. Table 1-5 contains a sample hypotheses about the relationship between a **physical trait (height) and a physiological factor** (pulse rate) that are listed in **Table 1-6: Finding Relationships.**

   Complete the table by adding your hypotheses.

TABLE 1-5
HYPOTHESES ABOUT RELATIONSHIPS
Example: The taller you are, the faster your pulse rate will be.
Your Hypotheses:

Check your hypotheses with your instructor before you continue.

TABLE 1-6 FINDING RELATIONSHIPS	
CHARACTERISTIC	METHOD OF INVESTIGATION
Height	Remove your shoes and stand against the wall. Use a meter stick to measure the distance from the floor to the top of your head (in centimeters).
Arm Length	Use a tape measure to measure the length of your arm from your shoulder joint to the tip of your middle finger (in centimeters).
Head Circumference	Use a tape measure to measure the distance around your head, just above your ears (in centimeters).
Pulse Rate	Place your index and middle fingers on your carotid artery (on either side of the neck). Using a watch that indicates seconds, count the total number of pulse beats for fifteen seconds. Multiply your answers by four to get the pulse rate per minute. Repeat the process twice more and then calculate the **average** number of pulse beats per minute.
Shoe Length	Measure the bottom of your shoe from the tip of the toe to the back of the heel (in centimeters).

ACTIVITY 6
DESIGNING THE EXPERIMENT AND COLLECTING DATA

1. **Work in groups of three to four students**.

   Choose **one hypotheses** to investigate from those developed by your group members.

2. **Form a plan** for getting the **data** you will need to test your hypotheses.

   The following supplies are available to conduct your experiment: **meter sticks** and **tape measures**.

   Collect your experimental data and record the results in **Table 1-7**.

TABLE 1-7 EXPERIMENTAL DATA		
SUBJECT		
1		
2		
3		
4		
5		
6		
7		
8		
9		
10		
11		
12		

Your fellow students will be happy to volunteer as test subjects in your experiment.

Hint:
How many pieces of information (data points) should you collect to feel confident about your conclusions? There are probably not enough people in your small laboratory group to make an adequate test sample. How many additional test subjects from the class will you need?

ACTIVITY 7   ORGANIZING AND SUMMARIZING RESULTS

To determine if there is a relationship between the two factors, **rearrange your data and copy it into the empty table below (Table 1-8)**.

Place the information in order from **smallest to largest** (shortest to tallest, lightest to heaviest, or smallest head to largest head).

TABLE 1-8 REARRANGED DATA		
SUBJECT		

ACTIVITY 8   GRAPHING YOUR RESULTS

1. Graphs provide a good visual representation of the relationships between the factors investigated in an experiment. **Bar graphs** and **line graphs** are frequently used to present scientific data.

   Figure 1-1 illustrates two different ways to present the same information.

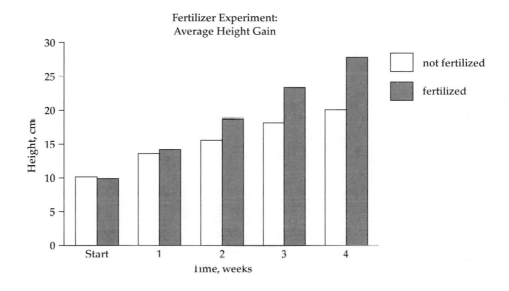

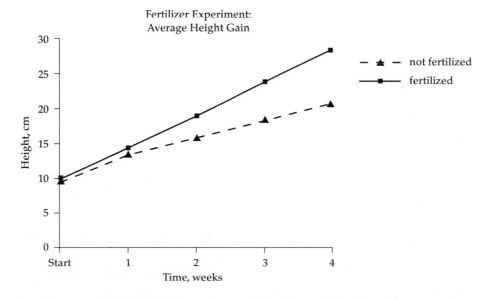

FIGURE 1-1. Comparison of Bar and Line Graphs

2. Look at the graphs in **Figure 1-1** and note the following **key points**:

    ■   The **horizontal** axis is referred to as the **X-axis**. The **vertical** axis is called the **Y-axis**.

        Each axis **must** have a title that clearly explains the numbers listed there.

    ■   Numbers on the X and Y axes must have an **equal interval** between them (for example, 5, 10, 15 but **not** 5, 10, 20, 50).

    ■   Lines or bars should be **large and easy to read**.

    ■   Numbers on the X and Y axes are chosen carefully to make **best use** of the space available.

    ■   Each graph should have a **title** that describes the subject matter being graphed.

    ■   It is not permissible to extend lines or bars **outside the margins** of the graph. Adjust the graph scale to make the data fit comfortably.

3. On the graph paper in **Figure 1-2, plot a graph of your experimental results**.

    Plot one characteristic on the **vertical (Y) axis** and the other on the **horizontal (X) axis**.

    Whenever possible, plot the **first characteristic mentioned in your hypotheses** on the **X- (horizontal) axis** and the **second** characteristic on the **Y- (vertical) axis**.

4. **Discuss the results** with your other group members.

    Write a conclusion based on your hypotheses and collected data. Support your conclusion by mentioning facts collected during your experiment.

Did You Know?

All the information, concepts, and relationships you will read about in your textbook were discovered and verified by the same scientific process you used in this laboratory exercise.

During this course, we will be summarizing hundreds of years of study and experimentation. Information presented on television, in newspapers, and in magazines often has not been confirmed by this same careful process and may not be correct.

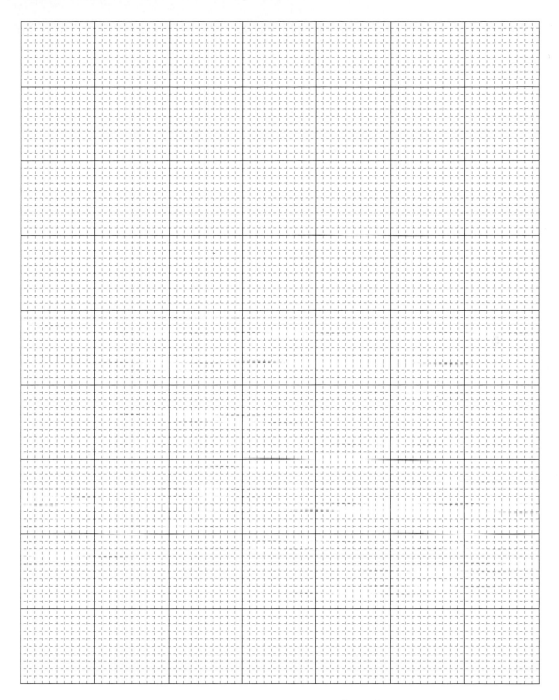

FIGURE 1-2. Experimental Results

ACTIVITY 9   INTERPRETING DATA USING SIMPLE STATISTICS

Preparation

Often, the data collected in an experiment is in a form that is not easily understandable. Measurements have been established to make it easier to interpret and draw conclusions from large collections of information. We're all familiar with the U.S. Census, which collects huge amounts of information on family size, income, housing conditions, population distributions, and so on.

Simple statistical analysis can reduce the data and convert it to a useable form. A similar approach is used when analyzing the results of large experiments (such as evaluating the effectiveness of new medications or airbags in automobiles). The most commonly used statistical measures are the mean, the mode, the median, the range, and the standard deviation.

The **mean** is the **average** of a set of numbers. The mean is equal to the **sum of all the numbers in the set divided by the sample size**. For example, to find the average pulse rate of a group of ten students, you would add the pulse values for each of the students together and then divide the answer by the number of students (10).

The mean of a group of numbers often doesn't really give you the information you need to correctly interpret the data. **For example, the following two sets of numbers have exactly the same mean, but the spread (dispersion) of numbers is quite different.**

Set 1: 39, 38, 38, 40, 40 mean = 39

Set 2: 3, 29, 25, 38, 100 mean = 39

The **mode** is the **most frequently occurring number** in a set. The mode represents the most common response and, therefore, can be used as a prediction to determine market response (for example, which car model will sell the best in a specific area of the United States).

The **median** is the **middle** number of a set when they are arranged in either **ascending or descending order**. If your income level is above the median, for example, your salary is in the upper 50% of salaries being compared.

The **range** is the **difference between the largest and smallest value** in the set. For example, the difference between the number of yards gained by the best and worst running backs in the National Football League.

To demonstrate how applying different statistical measures changes the meaning of results, consider the following set of 20 biology exam scores in **Table 1-9**.

TABLE 1-9 BIOLOGY EXAM SCORES			
STUDENT NUMBER	SCORE: EXAM 1	STUDENT NUMBER	SCORE: EXAM 1
1	90	11	88
2	94	12	54
3	80	13	32
4	82	14	47
5	91	15	25
6	46	16	56
7	97	17	59
8	96	18	60
9	87	19	87
10	84	20	86

Mean = 72.05

Mode = 87

Median = 83

1.   What is the **range** of exam scores in Table 1-9?

2.   If your **exam score was 80**, was your score in the top 50% of the class? **Explain your answer**.

3.   In this situation, is the **mean** a good representation of the class scores? **Why or why not?**

4.   While watching television last night, you saw an advertisement for Lose-Fast Weight Control Pills. The 12 women shown lost an **average** of 30 pounds while taking the pills. What additional **statistical** information would you need to make an informed decision whether or not to purchase this product?

Check your answers with your instructor before you continue.

Self Test

For each sentence below, enter the letter of the correct step of the scientific method.

a. Test hypotheses (by experiment or observation)

b. State hypotheses

c. State results (facts only)

1. _____ Seeds will grow faster if you fertilize the ground before you plant them.

2. _____ In an experiment, 70 of 80 household cockroaches were attracted to peanut butter.

3. _____ Tanya grew bacteria from her mouth on special plates in the laboratory. She placed drops of different mouthwashes on each plate.

4. _____ Kevin designed a survey to determine how many of his classmates had dimples on their chins and how many did not.

5. _____ Plants grown under red light will grow faster than those under white light.

6. _____ If acid rain affects plants in a particular lake, it might also affect small animals that live in the same water.

7. _____ Maria's experiment showed that chicken eggshells are more resistant to crushing when the hens are fed extra calcium.

Identify the graphing mistakes in **Figures 1-3, 1-4, and 1-5**:

8.

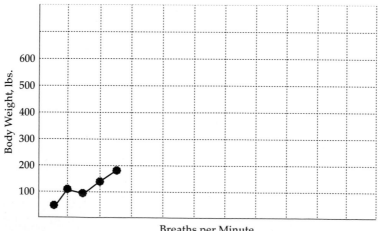

FIGURE 1-3. Sample Graph One

9.

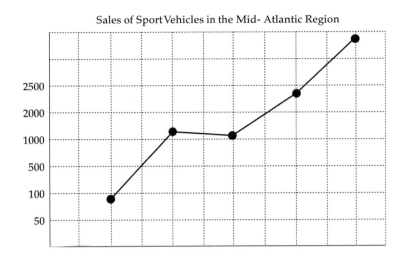

FIGURE 1-4. Sample Graph Two

10.

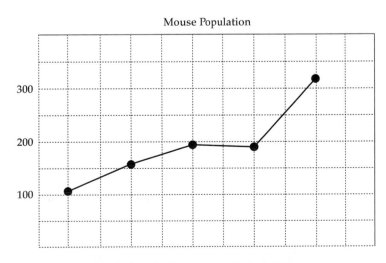

FIGURE 1-5. Sample Graph Three

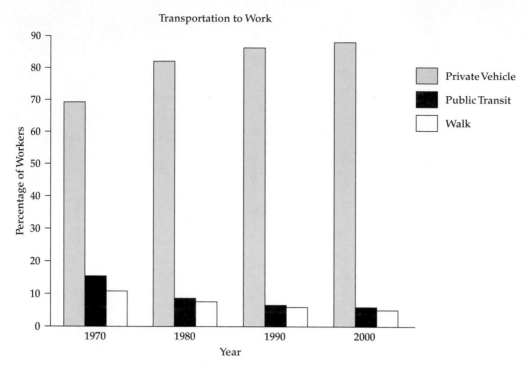

FIGURE 1-6.   Percentage of Workers Using Various
Modes of Transportation

Answer the following questions in reference to the graph in **Figure 1-6**.

11.   (**Circle one answer**.) Between 1970 and 2000, the percent of workers who
walked to work **increased/decreased/remained the same**.

12.   Between 1970 and 1980, the number of people who drove cars to work increased
by _____ %.

13.   In 2000, what percent of workers drove cars to work? _____.

14.   The number of workers who used public transportation dropped by 7%
between 1970 and _____.

EXERCISE
2

Interdependence Among Organisms

Objectives
After completing this exercise, you should be able to:
- explain the ways in which energy and nutrients are transferred through food chains
- discuss the connection between starch production and the process of photosynthesis
- demonstrate awareness of factors that can affect the accuracy of experimental results
- discuss factors that affect the ability of materials to decompose
- draw conclusions that are supported by experimental data
- apply your knowledge of photosynthesis and decomposition to real-life situations.

CONTENT FOCUS

Animals depend on plants for the energy to survive. Animals are part of food chains that begin with **producers** (such as plants) and continue through several levels of **consumers** such as humans, other animals, and many microorganisms.

The energy that makes animal life possible is obtained from producers. This familiar pattern is repeated in ecosystems all over the world. Even though deserts and rain forests have different types of producers and consumers, the pattern remains the same.

Decomposers (such as bacteria and fungi) are special types of consumers that feed on the remains of other organisms. During this process, raw materials are recycled and become available to producers.

21

The food chains described above illustrate an important biological principle—all organisms are linked together for survival and, therefore, are **interdependent**. Humans cannot separate themselves from these interactions. The more we understand about how living organisms interact, the better prepared we will be to ensure that these cycles continue.

The following cycle (**Figure 2-1**) shows how energy and nutrients are transferred from producers to consumers.

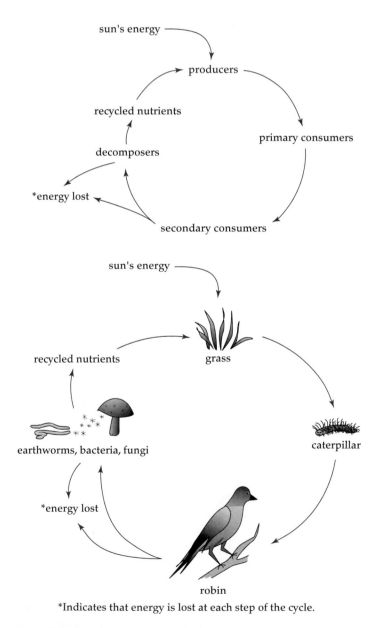

*Indicates that energy is lost at each step of the cycle.

FIGURE 2-1.   Transfer of Energy and Nutrients in a Food Chain

Where do producers get their food? They depend on food made by photosynthesis, which is powered by energy from the sun. Photosynthesis makes use of light energy to manufacture food molecules from carbon dioxide and water. Photosynthesis is the most important chemical process on earth. Producers have the green pigment **chlorophyll**, which **captures the light energy** used in photosynthesis.

During photosynthesis, producers **remove carbon dioxide** from the atmosphere and use it to form **glucose**. The process can be summarized with this equation:

$$CO_2 + H_2O \xrightarrow[\textit{chlorophyll}]{\text{sun's energy}} \text{glucose} + O_2$$

The food molecules that algae, green plants, and other producers make during photosynthesis form the **base of a food chain** that includes almost all living organisms on this planet (plants, animals, protists, fungi, and bacteria).

Excess energy is stored for later use. Plants **store** energy obtained through photosynthesis in the form of **starch**. To determine whether photosynthesis occurs in leaves, you will test for the presence of starch.

ACTIVITY 1   TESTING LEAVES FOR STARCH

1. Working in groups, get the following supplies: one or two freshly picked begonia leaves, a dropper bottle of iodine, a petri dish, and some paper towels.

   On your laboratory table, you will also find **an electric hot plate, two 400-ml beakers, a pair of long forceps, safety glasses, and a large beaker.**

Note:
More than one group will be using the hot plate set up at the same time—so you MUST coordinate your activities accordingly!

2. **Set up the following equipment**:

   Pour **300 ml of tap water** into the large beaker and set the hot plate on **high**. Heat the water to **boiling**.

   Boiling the leaves **softens the supporting walls of the leaf cells** so that chemical testing of the cell contents will be possible.

3. Fill **one of the 400-ml beakers** with **250 ml of tap water** and set it aside to use in **Step 9** of the procedure.

4. **Using the long forceps**, place **all the leaves** in the boiling water for **five minutes**.

   Reduce the heat setting to maintain a **low boil**.

5. While the leaves are boiling, **go to the supply area and put 125 ml of ethyl alcohol into the second 400-ml beaker**.

6. **Using the long forceps and being careful not to burn yourself**, remove the leaves from the boiling water.

   Place **all** the boiled leaves into the beaker of alcohol.

Caution!

Eye protection is required for the following procedures!

---

7. **Wearing eye protection, carefully** place the beaker of alcohol into the larger beaker of boiling water (see **Figure 2-2**).

8. Boil the leaves in the alcohol for **about five minutes**.

   (Don't forget to keep the heat setting at a **low** boil.)

   Continue boiling the leaves until they **lose their green color**.

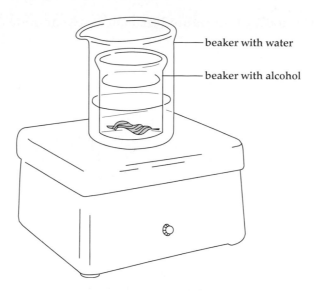

FIGURE 2-2. Setup for Starch Experiment

Note:

Read these directions COMPLETELY before proceeding.

---

9. **Wearing eye protection, use the long forceps to carefully** remove the leaves from the hot alcohol.

   As you remove each leaf from the alcohol, **dip it in the beaker of tap water** prepared in **Step 3**. This will remove the excess alcohol.

   Without tearing the leaf, blot it dry with paper towels and place one **leaf** in a petri dish **for each group**.

   Carefully, spread the leaf out as much as possible.

Note:

At this point, you will go back into your ORIGINAL groups to complete the experiment.

---

10. **Soak** the leaf with **iodine** (see **Figure 2-3**). Wait for **two minutes** and observe the leaf.

FIGURE 2-3. Starch Testing

Record your results:

11. Iodine is an **indicator** for the presence of starch. If starch is present, the iodine will **change color from reddish-brown to black**.

(**Circle one answer**.) In the iodine test for starch, my leaf result was **positive/negative**.

What **percent** of your leaf contained starch? (Make an **estimate**.) _____

> ### Note:
> **Save your leaf for use in later experiments.**

✔ Comprehension Check

1. A positive **iodine** test shows that _____ is present in a leaf.

2. A positive **iodine** test shows that the process of _____ has taken place in that leaf.

3. Rewrite this false statement to make it true.

   The **starch** in a leaf changes color when mixed with iodine.

ACTIVITY 2   EXPERIMENTS WITH THE STARCH TEST

Preparation

Two types of plants are available in your laboratory:

a. Plants with variegated (multicolored) leaves
b. Plants with tape covering part of each leaf

Using the information gained from **Activity 1**, form a hypotheses **concerning the presence of starch in leaves from only one of these plant types**. You will **design and perform** an experiment to test your hypotheses.

1. In the space below, **enter your leaf/starch hypotheses**.

STARCH HYPOTHESES

Check your hypotheses with your instructor before you continue.

2. **Carefully, remove** the alcohol beaker from the boiling water. **Add water** to the large beaker until it reaches the **300-ml level**.

Caution!
Both beakers are HOT. Be careful not to touch the hot glass.

3. **Perform** the experiment according to the instructions in **Activity 1** and record your results below. Remember that results **include only facts, not opinions!**

RESULTS OF STARCH EXPERIMENT

4. Draw a picture of your experimental leaf. In your drawing, **shade in the area(s) in which starch was found**.

PRESENCE OF STARCH IN EXPERIMENTAL LEAF

5. Which section of the leaf was the **control** for this experiment? **Explain** your answer.

6. Were your results similar to those of Activity 1? Why or why not?

7. **(Circle one answer.)** The results of this experiment **supported/did not support** my hypotheses.

8. **Write a conclusion based on your hypotheses and collected data. List facts collected** during your experiment to substantiate your conclusion.

✓ Comprehension Check

1. For those areas of the leaf in which you observed a **negative iodine test**, what can you say about the process of **photosynthesis?**

2. Can a plant store energy in its leaves in the **absence of light?** Explain your answer by mentioning **facts collected** during your experiment.

3. Suppose you placed a plant that contained starch in the dark and left it there for a week. At the end of the week, what results would an iodine test show? **Explain your answer**.

4. During the Gulf War, burning oil wells filled the sky with smoke and soot. Sunlight was blocked for several months. How might have this affected plants in the region? **Explain your answer**.

Check your answers with your instructor before you continue.

ACTIVITY 3   FORMING HYPOTHESES:
WHAT DECOMPOSES IN A LANDFILL?

Preparation

For ecosystems to continue functioning, raw materials and nutrients must continually be recycled back to producers.

Humans dispose of their wastes into the environment. Every man, woman, and child in the United States generates **four pounds of trash a day**. Most of us have only vague ideas—and often misconceptions—about what happens to trash after it leaves our houses. Much of the **180 million tons of trash** generated annually finds its way to landfills. Recycling of these materials depends on the action of decomposers present in the soil used to cover the trash.

The trash that goes to landfills includes more than just food wastes. Household items and packaging materials occupy a significant portion of landfill space. You will be performing an experiment to determine the **effect of decomposers** on commonly discarded items.

1. You will experiment with **eight materials** that are commonly found in landfills:

   Wood chips   Styrofoam packing material
   Popcorn   Cornstarch packing material
   Paper bags   Newspaper
   Aluminum foil   Plastic bags

2. For each material, **form a hypotheses about whether the material will decompose in your community landfill**.

3. **Record** your hypotheses in **Table 2-1**. These are your **personal hypotheses**. They do not have to agree with the hypotheses of other group members.

TABLE 2-1	
HYPOTHESES FOR DEGRADABLE AND NONDEGRADABLE MATERIALS	
DEGRADABLE	NONDEGRADABLE

ACTIVITY 4   CONSTRUCTING A LANDFILL

Perform the following experiment to determine whether different materials will decompose.

Note:
Read these procedures COMPLETELY before you begin.

1. You will perform this experiment in **groups. Each group will construct four land-fills.**

2. Each landfill will contain the following items:

Group 1:	Jar 1 (dry conditions)	Styrofoam packing material (3 pieces) Paper bags (3 pieces)
	Jar 2 (moist conditions)	Styrofoam packing material (3 pieces) Paper bags (3 pieces)
	Jar 3 (dry conditions)	Popcorn (3 pieces) Aluminum foil (3 pieces)
	Jar 4 (moist conditions)	Popcorn (3 pieces) Aluminum foil (3 pieces)
Group 2:	Jar 1 (dry conditions)	Cornstarch packing material (3 pieces) Newspaper (3 pieces)
	Jar 2 (moist conditions)	Cornstarch packing material (3 pieces) Newspaper (3 pieces)
	Jar 3 (dry conditions)	Wood chips (3 pieces) Plastic bags (3 pieces)
	Jar 4 (moist conditions)	Wood chips (3 pieces) Plastic bags (3 pieces)

3. **You must coordinate with another laboratory group**. These students will act as your data collection "partners." You **must** incorporate your partner group's data into your results.

4. Get the following supplies: **four plastic jars, four lids, and labeling tape**.

    Make a label for each jar similar to the sample label: **Jar 1 (dry)**

5. When all the jars have been labeled, get the following: **plastic spoons, metric rulers, one large beaker filled with soil**, and **six pieces of each of the four materials that were assigned to your group**.

6. **Measure** each piece of packaging material to get an **estimate of its surface area**.

    Record the **length and width** for each piece of packing material **in centimeters**. **Round** off the measurement to the nearest **tenth of a centimeter**. If the item is **not flat (such as the popcorn)**, consider the **thickness to be the width**.

7. **Calculate the approximate surface area** of each item:

$$\text{surface area} = \text{length} \times \text{width (or thickness)}$$

Average the surface area measurements for each material and **record** the results in **Table 2-2**.

Record the initial surface area measurements from your partner group with your own data in **Table 2-2**. This will complete the first two columns in **Table 2-2**.

TABLE 2-2 RECORDED DATA				
ITEM	INITIAL SURFACE AREA. AVERAGE VALUE FOR DRY SAMPLES	INITIAL SURFACE AREA: AVERAGE VALUE FOR MOIST SAMPLES	FINAL SURFACE AREA: AVERAGE VALUE FOR DRY SAMPLES	FINAL SURFACE AREA: AVERAGE VALUE FOR MOIST SAMPLES
Wood chips				
Popcorn				
Paper bags				
Pluminum foil				
Styrofoam packing material				
Cornstarch packing material				
Newspaper				
Plastic bags				

8. To construct your landfills, place a **layer of soil in the bottom of each jar**, spreading it evenly.

    Make **two duplicates** of each landfill (one **moist** and one **dry**). Make the two duplicates as similar as possible.

    In each landfill, place **two different materials** in layers, according to the instructions on the previous page.

    Lay each piece **as flat as possible** in the landfill container. **Separate each piece** with a layer of soil.

    It's fine to have more than one item per layer, as long as each item has **soil on all four sides of the item**.

9. **As you construct your landfills**, sprinkle **each layer** of your **two moist landfills** lightly with water to simulate rain.

    Keep adding water until you are satisfied with the level of moisture in the soil.

> ### Note:
> **The landfill should be moist but not soggy. Add the same amount of water to each "moist" landfill.**

10. Form a hypotheses: **will materials decompose equally well in dry and moist conditions?**

    For those items you hypothesized would be degradable, **return to Table 2-1** and write "**D**" after each item you predict will decompose under **dry** conditions and "**M**" after each item that you predict will decompose under **moist** conditions. It is perfectly acceptable to have **both "D" and "M"** listed for the same item.

11. **Cover the top layer** with soil, leaving about half an inch of space at the top of the jar. **Cover** your landfill with the appropriate lid. Dry off your jars and **attach a tape label identifying your group** (**DO NOT** write on the jar itself). Place the jars in a designated storage area.

12. Your landfill will "**incubate**" for **four weeks**.

    On the **fourth week**, you will take the landfill apart and **remeasure** your buried materials to determine the extent of decomposition.

13. When you take your landfill apart, **you will record your measurements** in **Table 2-2**.

    Include the results from your partner group with your own data in Table 2-2.

ACTIVITY 5   COMPLETION OF LANDFILL EXPERIMENT

1. From the storage area, retrieve the **four landfill containers** prepared by your laboratory group.

   Get a **large sheet of paper. Cover** your work area with the paper.

2. Empty the container onto the sheet of paper and **measure** the **final surface area** of each piece of material remaining. For measuring, follow the instructions in **Activity 2**.

3. **Record** the measurements for **all four containers** in the appropriate columns of **Table 2-2**.

4. **Record** the measurements taken by your **partner group** with your own data in **Table 2-2**.

5. Using the information collected, complete the **Report on Decomposition of Materials Commonly Found in Landfills** (Appendix II).

Self Test

1. An organism that is capable of producing its own food through photosynthesis is called a(an) _____.

2. List **three** examples of organisms from question 1 that were **not mentioned** in this laboratory exercise.

   a.

   b.

   c.

3. If an animal feeds on an organism that performs photosynthesis, it would be called a(an) _____ in a food chain.

4. If an insect eats a plant, and then a fish comes along and eats that same insect, the fish would be called a(an) _____ in a food chain.

5.   Draw a food chain for a vacant city lot that has **weeds, rats, cats**, and **crickets**.

      Label each member of the food chain according to whether it is a producer, primary consumer, and so on.

6.   If one of the cats died while it was in the vacant lot, what would happen **to that cat?**

7.   Give an example of a part of a plant in which starch is stored, but where photosynthesis **does not** take place.

8. *Challenge Question!* A student did an experiment to test the hypotheses that plants cannot grow in cold temperatures. The student used two groups of identical plants, with 10 plants in each group. One group was placed in the **refrigerator**, and the other on the **windowsill** in the classroom. Both groups were planted in the same soil and received exactly the same amount of water and fertilizer. The student checked on the plants twice a week for four weeks. At the end of the four weeks, the refrigerator plants were all dead, but the windowsill plants were fine. From the results of this experiment, the student concluded that the cold killed the plants.

Do you agree with this conclusion? **Explain your answer**. Support your explanation with facts about the **experimental design**.

Hint:
Somehow, this question must be connected to the material you covered in lab today!

Windows to a Microscopic World

Objectives

After completing this exercise, you should be able to:

- identify the parts of the compound and dissecting microscopes and explain their functions
- choose the correct type of microscope for viewing different specimens
- focus the compound microscope using the scanning, low-, and high-power lenses
- prepare a wet mount slide
- correct viewing problems that commonly occur when using the compound microscope
- accurately describe specimens viewed through the dissecting and compound microscopes
- use the microscope to test a hypothesis.

CONTENT FOCUS

All organisms, large and small, make valuable contributions to ecosystem function. Many plants and animals are large and easy to see, but many important organisms are too small to be seen without the assistance of magnifying lenses.

You will have an opportunity to get a close look at the anatomy and behavior of some interesting organisms that live around you unnoticed. Microscopes work like magnifying glasses to give you a closer look at these small organisms.

Different types of microscopes can be used for different purposes. During this exercise, you will be learning to use a **dissecting microscope** to examine larger objects and a **compound microscope** to view smaller specimens.

ACTIVITY 1

LEARNING TO USE THE DISSECTING MICROSCOPE

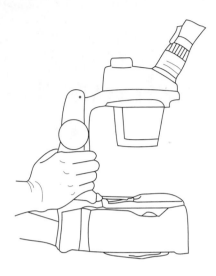

> ### *Caution!*
>
> **Dissecting microscopes may have several parts that are not permanently attached! Always carry the scope upright with one hand under the base and the other hand on the arm.**

1. Work in groups of **two students**. Get the following supplies: a **dissecting micro-scope** and a **penny**.

2. Place the penny on the microscope **stage**. Locate the stage by referring to the diagram in **Figure 3-1**.

    Turn the penny so that the Lincoln Memorial is facing you. **Adjust the light** until the penny is brightly illuminated.

3. Set the **magnification control knob**, located on the top or the side of the **head**, to the **lowest setting**.

4. Turn the **focusing knob**, located on the microscope **arm**, until the head is as close to the stage as possible.

    Look through the **ocular lenses (eyepiece)** at the penny. **Turn the focusing knob** until the image of the penny is sharp and clear (the head will be moving **away** from the stage).

    Is Lincoln sitting in the Lincoln Memorial in your penny? _____

5. **While looking through the eyepiece, gradually turn the magnification control knob**. The change in image size will resemble the zoom action of a camera.

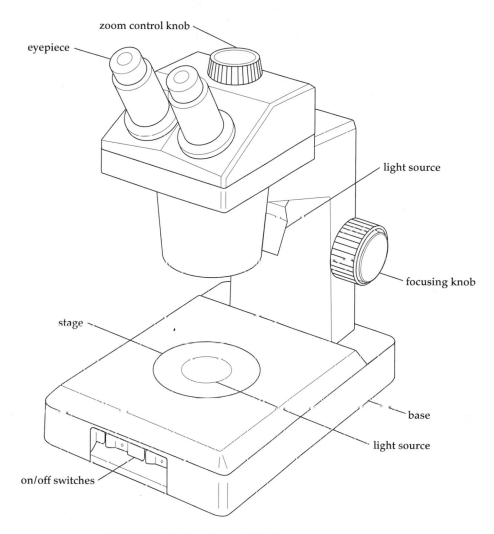

FIGURE 3-1. Parts of the Dissecting Microscope

ACTIVITY 2

MAKING OBSERVATIONS WITH THE DISSECTING MICROSCOPE

1. Work in groups. Get the following supplies: **a small glass bowl, a pipette, and a blunt metal probe**.

2. Use the pipette to remove a **planaria** from the culture container and place it in your bowl. Locate the worm of your choice and draw it into the tip of the pipette (as shown in **Figure 3-2**). When the planaria feels the water current, it will probably form itself into a small protective ball, and it will be easy to draw it into the pipette.

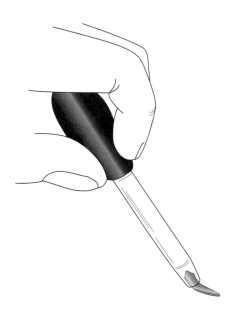

> ## *Caution!*
>
> **Don't suck the planaria too far into the pipette. It will attach to the inside of the pipette and you won't be able to get it out.**
>
> **If additional water is needed in your glass dish, add pond water. Don't use tap water or distilled water.**

FIGURE 3-2. Drawing Planaria into Pipette

3. **Observe your worm** under the **dissecting microscope**. Use a medium or low light level.

   Observe carefully through the microscope for at least **two minutes**.

   Does the worm swim or crawl? _____

   Describe the movement of the worm's **muscles** as it moves forward.

4. **Draw a picture** of your worm. Make it **large** and **clear. Label** the **head** and the **eyes**.

DRAWING OF PLANARIA

5. Planaria are common inhabitants of freshwater ponds and streams. On the basis of your **observations**, do you think they are **producers** or **consumers? Explain** your answer.

6. Hold the **blunt probe motionless**, directly in the path of a moving worm. How does the worm respond?

7. Touch the head end **gently** with the blunt probe. How does the worm respond?

8. Touch the planaria **gently** on several other body parts. Record your observations.

9. How was the **touch response** on other body parts **similar or different** to the response you observed when the worm was touched on the head end?

Check your answers with your instructor before you continue.

ACTIVITY 3   GETTING FAMILIAR WITH THE COMPOUND MICROSCOPE

1. Work in groups. Get a **compound microscope**.

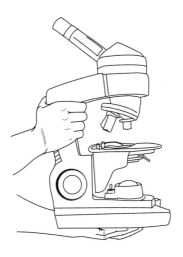

> ### *Caution!*
> **Always carry the scope upright with one hand under the base and the other hand on the arm.**

2. Plug the microscope into an electrical outlet and **turn the light on**. The light switch is located on the **base** of the microscope.

3. The compound microscope consists of a system of **optics** (lenses and mirrors) and focusing controls.

   The base and the arm support a **body tube** that **houses the lenses** that magnify the image.

4. At the top of the scope is the **ocular lens (eyepiece)**. The ocular is only one of a series of lenses that magnify the image. The ocular lens makes the image **ten times larger** than life size (abbreviated 10×).

   Look through the ocular lens. You will see a black pointer. Rotate the eyepiece **while looking through the lens**.

   What happens to the pointer? _____

5. At the bottom of the body tube is a revolving **nosepiece**. The lenses that screw into the nosepiece are called **objective lenses**.

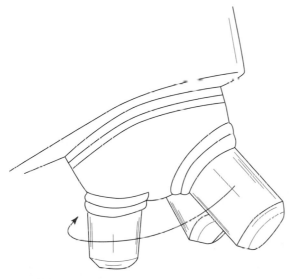

Turn the nosepiece until you hear one of the objective lenses **click into position** (see **Figure 3-3**).

When an objective lens clicks into position, it is in the proper alignment for light to pass from the light source, through the objective, through the ocular lens, and into the viewer's eye.

Turn the nosepiece again to bring a different objective lens into position.

FIGURE 3-3. Revolving Nosepiece and Objective Lenses

6. **How will you know** when you have placed the objective lens in the proper position?

7. Note that each objective lens is of a different length. Each of the lenses has a different magnifying power. The **shortest lens** has the **lowest magnifying power** and the **longest lens** has the **highest**.

   Since light from the specimen passes through **both** an objective lens **and** the ocular lens, the **total magnification** of the image is the result of the objective lens magnification **multiplied** by the ocular lens magnification.

8. Using the information on the previous page, **complete the magnification table**.

Length of Objective Lens	Magnification of Objective Lens	Magnification of Ocular Lens (Eyepiece)	Total Magnification of Specimen
Short	4×		
Medium	10×		
Long	40×		

9. On both sides of the arm, you will see the **focus knobs**. The larger ring is the **coarse focus knob** and the smaller ring is the **fine focus knob**.

   Turn the microscope so that you are looking at it **from the side**. Looking at the lenses and stage, turn the **coarse focus knob** all the way in one direction and then reverse the process. **What happens?**

   Rapidly turn the **fine focus knob** all the way in one direction and then reverse the process. **What happens?**

10. The **mechanical stage** is a moveable platform designed to hold a microscope slide.

    The distance between the stage and the objective lens is adjusted by using the coarse and fine focus knobs.

    Two control knobs move the mechanical stage. One moves the stage left and right; the other moves the stage forward and back. **Always move the stage using the control knobs**. Don't attempt to move the slide using your fingers.

11. In the center of the stage, you will see the glass of the **condenser lens**, which focuses the light on the specimen.

12. Directly above the light source, you will see a lever that moves from left to right. This lever is connected to the **iris diaphragm**. Looking at the condenser lens, move the **iris diaphragm lever** all the way to the **left. What happens?**

13. Looking at the side of the microscope, move the **iris diaphragm lever** all the way to the **right. What happens** now?

ACTIVITY 4   PARTS OF THE COMPOUND MICROSCOPE

Now that you have some experience with the parts of the microscope and their functions, **label the parts** of the microscope on **Figure 3-4**.

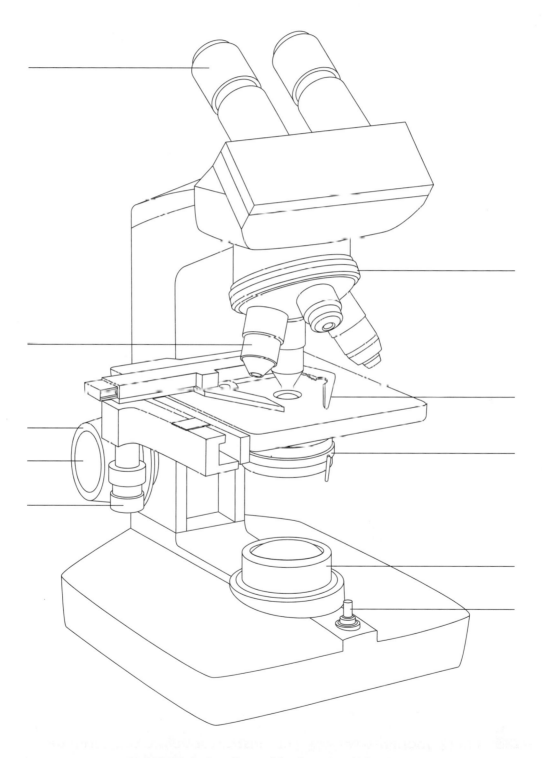

FIGURE 3-4. Parts of the Compound Microscope

ACTIVITY 5
LEARNING TO USE THE COMPOUND MICROSCOPE

1. Work in groups. Get the following supplies: a **prepared slide** labeled **letter "e"** and a piece of **lens paper**.

2. **Clean the letter "e" slide** with the lens paper.

    Using the lens paper, **clean the ocular lens**.

    Be sure that the letter "e" is **right-side up**, in its **normal reading position** before you place it on the stage.

3. Move the **iris diaphragm** lever to about the middle position. Click the **scanning (lowest power) objective lens** into place.

4. Turn the **coarse focus** until the objective lens is as close to the slide as it will go.

    While looking through the eyepiece, slowly raise the objective lens to bring the letter "e" into focus.

✔ Comprehension Check

1. If the letter "e" is not lighted brightly enough for you, **what should you do?**

2. If the light is too bright, causing glare and eyestrain, **what should you do?**

3. What is the **total magnification** of the letter "e" image? **Explain** your answer.

Check your answers with your instructor before you continue.

5. Draw the letter "e" **exactly** as it appears under low power. Make it **large** and **clear**.

LETTER "e"

6. How is the **orientation** of the letter "e" as seen through the microscope **different** from the way an "e" **normally** appears? List **two** differences.

7. **While looking through the eyepiece**, move the stage to the **left**.

   What direction does the image appear to move? _____

   While looking through the eyepiece, move the stage **away from you**.

   What direction does the image appear to move? _____

If you wanted to **center** the letter "e" in this drawing, which **two directions** would you have to move the stage?

_____ and _____

8. **Centering the specimen is** *absolutely necessary* **before you can change to a higher power objective!** Do you know why?

   The reason is simple. The more powerful the magnifying lens, the smaller area you see, but you see that small area in greater detail. The area you can see at one time is called the **field of view**. So you could say that the **higher** the power of the objective, the **smaller** the field of view.

9. Center the letter "e" image, rotate the nosepiece, and **click in the 10× objective lens** (also called the **low power objective**).

Note:

Do not move the coarse adjustment knob!

Looking through the eyepiece, can you still see the letter "e"? _____

Is the letter "e" exactly in the center of the field of view? _____

If not, **move the slide slightly** to center the image.

Is the letter "e" sharp and clear? _____

If not, **gradually adjust the fine focus knob** until the problem is corrected.

Do you have enough light? _____

If not, **gradually adjust the iris diaphragm**.

How has the **image of the letter "e" changed** from the way it looked using the 4× objective lens?

10. Repeat the instructions in **Step 9**, this time changing **from 10×** to the **high-power objective** (40×).

    Looking through the eyepiece, can you see the entire letter "e"? _____

11. **(Circle one answer.)** Your field of view on high power is **larger/smaller** than the field of view on 10×.

    How does the size of the field of view determine how much of the letter "e" you can see?

12. When using the **high-power objective** (40×), what is the **total magnification** of the letter "e" image? _____ **Show your work and explain your answer**.

 Check your answers with your instructor before you continue.

ACTIVITY 6

PREPARING TEMPORARY SLIDES— WET MOUNTS

Preparation

A wet mount is a method of preparing a slide that will be used only for a short time. Unlike the letter "e," which was permanently attached to the slide, a wet mount is made by placing the specimen into a drop of liquid on a slide. The specimen and water droplet are held in place by a coverslip.

Human Epithelial Cells (Cells from the Inside of Your Cheek):

1. Work in groups. Your group will make **two different wet mounts** of cheek cells.

   One will be made with a drop of **stain**. The second will be made by substituting a drop of **physiological saline** for the stain.

2. Get the following supplies: a **dropper bottle of iodine stain or physiological saline, a slide, a cover slip**, and a **clean toothpick**.

3. **Place a drop of iodine stain or a drop of saline near the center of a clean slide.**

4. **Gently** scrape the inside of your cheek with the **end of a toothpick.**

> ### *Caution!*
> If you scrape too hard, you will be examining blood cells instead of epithelial cells!

You have removed some of the cells that form a **protective covering** for the inside of the mouth. Like other epithelial cells, these are constantly being worn off and replaced by new cells of the same type.

5. **Spread the material from the toothpick** into the drop of water or stain. Add a coverslip.

   Put two microscopes together (yours and your partner's) on the laboratory counter so that you can view and compare the stained and unstained slides.

> ### *Hint:*
> Unstained cells are clear. They are only visible with very low light levels. If you think there are no cells on your slide, adjust the iris diaphragm!

6. **View both slides. Begin with the 4× objective.** Continue until you have located the cells using **all three objective lenses**.

7. **Draw a picture** of **one** cheek cell, **viewed on high power**. Make it **large and clear**.

HUMAN CHEEK CELL

8. Label the following cell structures (referred to as **organelles**) in your drawing of the cheek cells:

Organelles to Label	Function
Nucleus	directs all cell activities
Cell membrane	controls movement of materials in and out of the cell
Cytoplasm	jelly-like fluid found between the nucleus and cell membrane

9. Give an example of a **medical procedure** in which epithelial cells are scraped from another area of the body and examined microscopically.

10. Is there an advantage in **using a stain** to view cells microscopically? _____ If so, how is the stain helpful?

Daphnia—An Aquatic Organism

The water flea, daphnia, is a microscopic organism commonly found in ponds, lakes, and streams. Because it is small and transparent, living daphnia can be studied easily in the laboratory. Daphnia feed on microscopic food particles. Their five pairs of legs are modified into strainers that filter the food particles from the water. Daphnia, in turn, are an important part of the food chain. Many fishes and even larger aquatic animals feed on daphnia.

1. Work in groups. Get the following supplies: a **depression slide** and a **bottle of methyl cellulose** (the bottle may also be labeled **Protoslo®**).

2. Put a **small** drop of the methyl cellulose in the depression on your slide.

   Use the pipette in the culture jar to remove a **daphnia** from the container and place it in the depression on the slide. **You don't need a coverslip.**

3. Make your observations using the **scanning lens** of the microscope (**4×**).

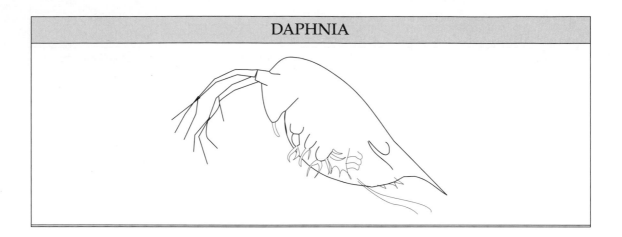

DAPHNIA

4. Add the following details to the daphnia outline below: **eye, heart, and intestine. Label the details you put in the drawing. Also, label the head and legs**.

ACTIVITY 7   USING THE MICROSCOPE TO ANSWER A QUESTION: WHAT DO AQUARIUM SNAILS EAT?

1. Work in groups. Go to your **classroom aquarium** and **observe the feeding activity** of the snails.

2. **Write a hypothesis**: What are the snails **eating?**

   Hypothesis:

Hint:
Don't forget to make sure your hypothesis is testable!

3. Develop an observation plan to **test your hypothesis**. List the steps of your plan below:

STEPS OF YOUR OBSERVATION PLAN

Check your hypothesis and procedures with your instructor before you continue.

4. **Carry out the steps** of your observation plan. Refer to **Figure 3-5** to help you identify organisms you observe.

   Record the results of your observations.

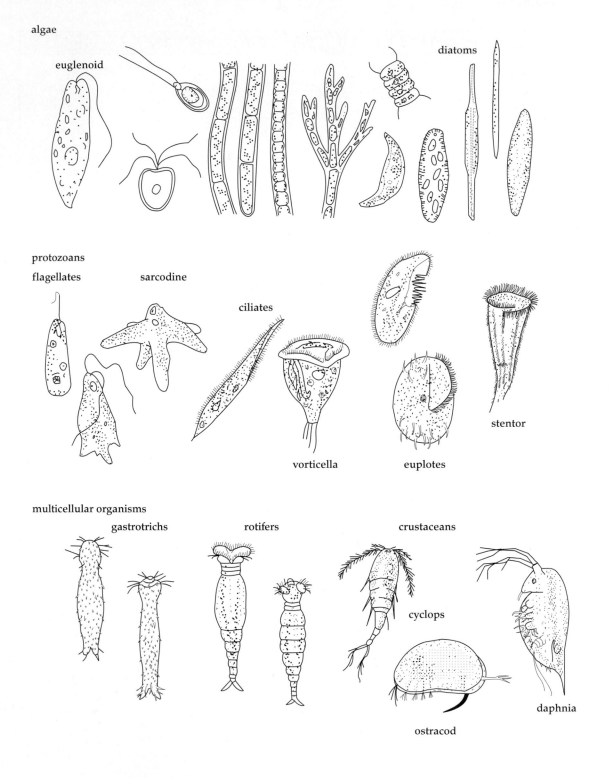

FIGURE 3-5. Organisms Commonly Found in Pond Water

OBSERVATION RESULTS

Comprehension Check

1. **Discuss** the results with your other group members. **(Circle one answer.)** My hypothesis **was/was not** supported.

   Write a conclusion based on your hypothesis and collected data. Support your conclusion by mentioning facts collected during your experiment.

Self Test

1. Complete the table by entering the appropriate part of the compound microscope or the correct function for the stated microscope part.

PARTS OF THE COMPOUND MICROSCOPE	FUNCTION
Ocular lens (eyepiece)	
Stage	
	Focuses light on the specimen
Nosepiece	
	Objective lens used to first locate a specimen
	Regulates the amount of light that passes through the specimen
Fine focus knob	
	Objective lens with the lowest magnifying power
	Objective lens with the highest magnifying power
Coarse focus knob	

2. What is the **total magnification** if the ocular lens is 15✕ and the objective lens is 20✕? **Show your work!**

3. List **three differences** between the dissecting microscope and the compound microscope:

    a.

    b.

    c.

4. **What type of microscope** (compound or dissecting) would you use to observe the following:

    _____ Cells from the lining of your stomach

    _____ A seashell you found on the beach

    _____ A cockroach you found in the kitchen

    _____ Mildew from your shower curtain

5. Suppose you were watching a daphnia under the microscope and noticed that it moved **toward you** and then **to your right**.

    Which direction(s) did the daphnia **actually** move? _____

    Explain your answer.

6.   Explain how you would **correct the following problems** experienced when using a microscope:

a. When changing magnification from 10× to 40×, the specimen disappears.

b. The field of view is too dark.

c. Your field of view is partially obscured by a dark area.

d. There is a fingerprint in your field of view.

e. There are many hollow, dark circles in your field of view.

Functions and Properties of Cells

Objectives

After completing this exercise, you should be able to:

- identify and explain the functions of the major cellular organelles
- explain the similarities and differences between plant and animal cells
- explain the concepts of diffusion and osmosis and why they are important to cell physiology
- use indicator chemicals to test for the movement of molecules and to determine the direction of diffusion
- explain the process of osmosis in living cells exposed to different extracellular solute concentrations
- apply your knowledge of cell structure and function to real-life situations.

CONTENT FOCUS

Typically, when you look at a plant or animal, it is easy to think of an organism as one large unit. When looking through the microscope, however, it becomes obvious that an organism is composed of trillions of tiny units called **cells**, which work cooperatively to carry out the functions that keep us alive. If the activity of your cells stopped, even for a few moments, death could quickly follow.

There are about 200 different types of cells in the human body. These cells can be quite different in shape, structure, and function, but they all have some basic characteristics in common.

Within each cell lies a collection of specialized structures called **organelles**, in which particular chemical activities take place. The various cell structures carry out activities that mirror the functions of our body organs. Most organelles are compartments bounded by membranes.

61

Learning about cell organelles will be more meaningful when you understand the role that each part plays in the life of a cell. **Table 4-1** gives a brief summary of the functions of some basic organelles.

TABLE 4-1 **SOME CELL ORGANELLES AND THEIR FUNCTIONS**	
ORGANELLE VISIBLE WITH THE LIGHT MICROSCOPE	FUNCTION
Cell wall	External support and protection; if present, located outside the cell membrane
Cell membrane	Surrounds the cytoplasm; barrier between the environment and the cell that controls the movement of materials in and out of the cell
Cytoplasm	Liquid/gel "filler" substance inside the cell membrane; all internal cell are suspended in the cytoplasm
Nucleus	Stores and transfers information (DNA and RNA) needed to control cell functions
Chloroplast	Structure in which photosynthesis takes place; contains chlorophyll and is green in color
ORGANELLE NOT VISIBLE WITH THE LIGHT MICROSCOPE	FUNCTION
Mitochondria	Location of aerobic cellular respiration; produces ATP energy
Ribosomes	Protein synthesis; ribosomes may be found free in the cytoplasm or attached to the endoplasmic reticulum (ER)
Endoplasmic reticulum (ER)	Membrane manufacturing complex
Rough endoplasmic reticulum (ER)	Ribosomes attached to ER membrane; protein synthesis
Golgi apparatus	Packaging of molecules for export out of the cell; most exported molecules are proteins

ACTIVITY 1

OBSERVING MOVEMENT INSIDE LIVING CELLS

1. Work in groups. Get **two compound microscopes** and set them up, side by side.

2. Make **two wet mounts**, one of **amoeba** and one of **elodea**. Place one wet mount on each microscope.

    Amoeba is a **one-celled organism** commonly found in pond water. **Elodea** is a **freshwater plant** frequently used in home aquariums.

3. **Make a drawing** of each organism, **labeling** all the cell parts you can identify.

    Use the information in **Table 4-1** to help you locate and identify the cellular organelles.

AMOEBA	ELODEA

4. Which organism(s) can carry on photosynthesis? _____

    What **evidence** did you see that led you to draw this **conclusion?**

5. Which organism(s) has a **cell wall?** _____

6. In which organism(s) are you able to observe the cytoplasm moving? _____

7. Some organisms can secure food by surrounding their prey with cell extensions. This process is called **phagocytosis** (shown in **Figure 4-1**).

    Which organism(s) do you think would be able to do this? _____

    Explain your answer.

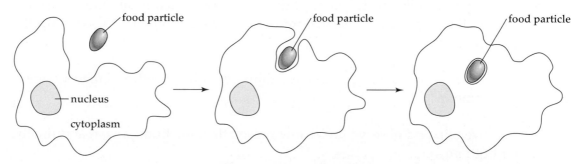

FIGURE 4-1.   Phagocytosis

8. Which **cell organelles** can be seen changing position in the moving cytoplasm of each organism?

    elodea _____

    amoeba _____

9. Which organism demonstrates **locomotion** (ability to move from place to place)?
    _____

    Write a **clear description** of the **movement pattern** of this organism.

   Check your answers with your instructor before you continue.

ACTIVITY 2   WATER MOLECULES IN MOTION

Preparation

You have just observed the movement of cytoplasm inside cells. Everywhere around us, molecules of liquids and gases are in constant, random movement.

Of course, these moving molecules are much too small to be observed directly, but we can demonstrate this activity by placing some colored powder into a drop of water.

1. Work in groups. Get the following supplies: **a slide, a coverslip, a dropper bottle of distilled water, a toothpick, and carmine powder.**

2. Place a drop of distilled water on the slide. **Carefully** add a tiny amount of carmine powder to the water drop. Use the **pointed tip** of the toothpick to scoop up the carmine powder.

    Imagine how much carmine powder you think you might need, and then scoop up half that amount.

    Add a coverslip and observe under low power, then medium power, and finally, under **high power**.

3. **Describe** the activity of the carmine particles in the water droplet.

4. Since carmine particles are **not** alive, what is **causing** the carmine particles to move?

   As was mentioned earlier, molecules of liquid are always in motion. This means water molecules everywhere are constantly moving and bumping into each other.

   As they move, they will also collide with anything floating in the water, such as the carmine particles.

5. **(Circle one answer.)** If we **warmed up** the carmine slide a little bit, the rate of molecular motion would be **faster/slower/stay the same.**

   Explain your answer.

You have just observed that molecules are in constant motion. Movement of molecules will allow us to explain **diffusion and osmosis,** two important cell processes.

The process of **diffusion** occurs whenever dissolved particles move from an area of **high concentration** (more of them) to nearby areas where they are **less concentrated.** The diffusion process is similar to a bumper car ride (see **Figure 4-2**). Cars are moving quickly and often collide. The force of the collision sends the cars shooting outward into an empty area of the bumper car rink. The more an area is crowded with bumper cars, the more collisions occur.

In this way, the cars are gradually dispersed from the crowded center of the rink (an area of **high** bumper car concentration) to the emptier fringe areas (areas of **lower** bumper car concentration). This is a good model of what happens when particles diffuse in the environment (see **Figure 4-2**).

Diffusing molecules always move outward from an area of high concentration into areas of lower concentration.

A similar process to diffusion occurs with **water** molecules in the environment. Cells are surrounded by a membrane that allows some substances to pass through but not others. This is referred to as a **selectively permeable membrane.**

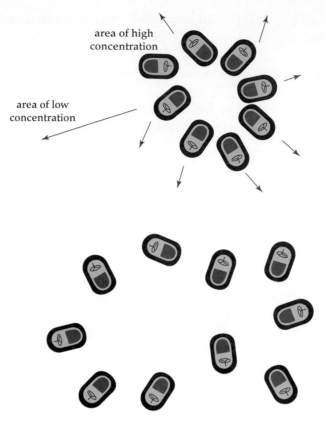

area of high
concentration

area of low
concentration

FIGURE 4-2. Bumper Car Model of Diffusion

Water molecules disperse through a **selectively permeable membrane** from an area of **high concentration** to an area of **lower concentration**. This process is given a different name, **osmosis**. We can demonstrate the processes of diffusion and osmosis in an experiment using **dialysis tubing**, an artificial membrane with many small pores.

ACTIVITY 3   DIFFUSION AND OSMOSIS

1. Work in groups. Get the following supplies: one plastic **jar, one** piece of **dialysis tubing, and two long pieces of thread**.

    Dialysis tubing is an **example** of a selectively permeable membrane. **Cell membranes** are also selectively permeable.

2. **Wet the piece of dialysis tubing** for a minute or two and open it as demonstrated by your instructor.

    Referring to **Figure 4-3**, construct a **dialysis tubing bag** by tying one end **tightly** with the thread. Wrap the thread several times around the tubing before tying to prevent leakage.

3.   Get a bottle of **stock solution.** The solution consists of **water** with a mixture of dissolved **salt** (NaCl), **sugar** (glucose), and **starch.**

Hint:
Starch tends to settle out on the bottom of the bottle. Make sure you shake the stock solution to mix completely before measuring.

Using a graduated cylinder, measure 20 **milliliters (ml)** of stock solution and pour it into your dialysis tubing bag.

Using the remaining piece of string, tightly **tie the open end** of the dialysis tubing.

FIGURE 4-3.   Dialysis Tubing Bag

4.   **Weigh your bag** according to the following directions, using the balance in your classroom:

   a. Pat the bag dry. Cut off any excess string.

   b. Make sure the **balance is set to zero.** (If you have any problems using the scale, consult your laboratory instructor.)

   c. Weigh the bag to **the nearest tenth of a gram.**

   Record the **initial weight** here: _____

5.   Fill the plastic jar about three quarters full with distilled water. Do not fill the jar with tap water. **Immerse** the dialysis tubing bag in the jar of distilled water.

6.   **Write a hypothesis.** What do you think will happen to the weight of the bag during the 20 minutes it sits in the jar of distilled water?

7. Set the jar aside for **at least 20 minutes.**

    While you are waiting, complete the Comprehension Check and then set up the controls for chemical testing in Activity 4.

    You will return to your dialysis experiment when the 20 minutes have elapsed.

✔ Comprehension Check

1. When someone uses a 10% glucose solution, what is actually in this solution? Fill in the blanks with the correct percentages (this must add up to 100%).

    _____ % glucose   _____ % water

2. **Figure 4-4** represents a dialysis tubing bag containing dissolved salt, sugar, and starch, surrounded by distilled water.

FIGURE 4-4. Dialysis Bag Surrounded by Distilled Water

(Circle one answer.)

The **highest** concentration of **salt** is inside the bag/outside the bag.

The **highest** concentration of **sugar** is inside the bag/outside the bag.

The **highest** concentration of **starch** is inside the bag/outside the bag.

The **highest** concentration of **water** is inside the bag/outside the bag.

Check your answers with your instructor before you continue.

ACTIVITY 4
CONTROLS FOR THE DIFFUSION AND OSMOSIS EXPERIMENT

Preparation

You will perform **three** different chemical tests to determine the results of your experiment. Each uses an **indicator chemical that changes color** if starch, salt, or sugar is present.

a. **Iodine Test for the presence of starch.** When added to a solution, iodine will turn **black** if starch is present. If no starch is present, the iodine will remain **reddish-brown** in color.

b. **Silver Nitrate Test for the presence of chloride ions.** When added to a solution that contains chloride ions, the silver nitrate will change color from **clear** to **cloudy white**. (Remember: salt is composed of sodium ions and chloride ions.)

c. **Benedict's Test for simple sugars.** When added to a solution that contains simple sugars, such as glucose, Benedict's solution will change color from **turquoise blue** to one of the following colors: **green, yellow, orange, or red.** Green represents the smallest amount of simple sugar and red the highest.

> **Note:**
>
> **It is important to remember that the starch, salt, and glucose are not changing in any way. It is the indicator chemicals that are changing color.**

1. Work in groups. Get a **large beaker.** Fill it **one-third full** with **tap water** and heat it to boiling.

   While you are waiting for the water to boil, set up the materials needed for chemical testing.

2. To set up your **control** tubes, get the following supplies: a **test tube rack, a ruler, a marking pencil, a test tube holder, and three test tubes.**

   Label the test tubes salt, sugar, and starch.

3. Using the ruler, place a **line** on each test tube **1 cm from the bottom** of the tube.

   Fill each of the three test tubes to the 1-cm line with stock solution.

4. To the test tube labeled **salt**, add **one dropper full** of **silver nitrate** solution.

Caution:
Avoid getting this solution on your skin.
---

5. Shake the tube gently to mix the contents. **Observe the color** of the solution. **Record your results** in the appropriate columns in **Table 4-2.**

Note:
Save all three of your control test tubes (salt, starch, and sugar) for later use.
---

6. To the test tube labeled starch, add **one dropper full** of **iodine** solution. Shake the tube gently to mix the contents.

   Observe the color of the solution. **Record** your results in **Table 4-2.**

7. To the test tube labeled sugar, add **one dropper full** of **Benedict's** solution. Shake the tube gently to mix the contents.

8. **Carefully** lower the test tube into the boiling water bath and allow it to remain for **two to five minutes.**

Caution:
Hot glass looks exactly like cold glass. Use a test tube holder to remove test tubes from the boiling water. Don't leave the test tube holder clamped on the test tube while in the boiling water. Hot metal looks like cold metal!
---

9. **Using a test tube holder**, remove the tube from the boiling water bath and place it in the test tube rack.

10. **Observe the color** of the solution. **Record** your results in **Table 4-2**.

T A B L E 4 - 2 RESULTS OF CHEMICAL TESTS				
INDICATOR SOLUTION	SHOWS THE PRESENCE OF	INITIAL COLOR	FINAL COLOR	RESULTS (+ OR −)
Silver Nitrate Test Control Tube				
Iodine Test Control Tube				
Benedict's Test Control Tube				
Silver Nitrate Test Experimental Tube				
Iodine Test Experimental Tube				
Benedict's Test Experimental Tube				

ACTIVITY 5   COMPLETING THE DIFFUSION AND OSMOSIS EXPERIMENT

1. **Remove** your dialysis tubing bag from the jar of water. To test for **osmosis**, follow the **same weighing directions** as in **Activity 3, Step 4, and weigh your bag again.**

2. Record the **final weight** here: _____

   Did the bag **gain or lose weight?** _____

   Was your hypothesis about the **weight change** correct? _____

3. Calculate the **percent change in weight** and record your answer.

   $$\text{percent change} = \frac{\text{final weight} - \text{initial weight}}{\text{initial weight}} \times 100$$

   percent weight change = _____ %

   What do you think **moved through the membrane** to cause this result? _____

4. To test for **diffusion**, get **three more test tubes. Label the test tubes salt, sugar, and starch.**

    Using the ruler, place a **line** on each test tube **1 cm from the bottom** of the tube.

    Fill each of the three test tubes to the **1-cm line** with **water from the plastic jar that held the dialysis tubing**.

    Test the water for salt, starch, and sugar using the same indicator chemicals you used when setting up the controls in **Activity 4**. Each test tube will contain **only** water from the jar and the indicator chemical.

    Do not add stock solution to these test tubes!

5. **Record** the results of all three tests in **Table 4-2**.

    According to your test results, **list the molecules** that were able to **diffuse** out of the dialysis bag.

6. If molecules are **expected** to move from **high concentration** to **low concentration**, why didn't **all** the molecules (salt, sugar, and starch) leave the dialysis tubing bag?

7. The ability to separate molecules from each other using a **selectively permeable** membrane is referred to as **dialysis**. Did the experiment you just finished illustrate this concept? **Explain your answer.**

Check your answers with your instructor before you continue.

ACTIVITY 6   OSMOSIS IN RED ONION CELLS

1. Work in groups. You will be making **two different wet mounts** of onion cells.

   One will be made with a drop of **distilled water**. The other will be made by substituting a drop of **20% saline** for the water.

   Get the following supplies: **a piece of onion bulb, a scalpel, forceps, a dropper bottle of distilled water, a dropper bottle of 20% saline solution, two slides, and two coverslips.**

2. Make wet mounts of the onion-skin cells, following the directions in **Figure 4-5**.

   With a marking pencil, **write** "S" on the right side of the slide to indicate saline solution.

Pull off a section of a **single leaf** from the **purple** skin on the **outer surface** of the onion bulb.

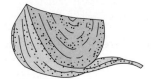

With forceps remove a piece of that thin **purple** skin and place it in the drop of **water** or **salt solution** on your slide.

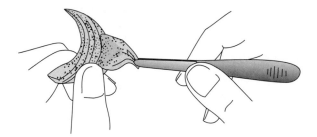

The piece of onion skin you remove **must be thin and transparent**. It will look like a small piece of purple plastic wrap.

Gently lower a coverslip over the specimen.

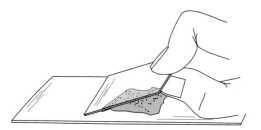

FIGURE 4-5.   Wet Mount of Onion Skin

3. Place **two microscopes** side by side and **compare** the two slides. **Start with the** 4× **scanning lens**. Center and focus the specimen. Change to **low**, and then to **high** power.

Note:

Many plant cells contain an organelle for water storage called a sap vacuole. In the plants you are examining, pigments in the leaf give the sap vacuole a pink color.

---

4. Wait **five minutes**, then check on both slides again.

    Draw a large picture showing **one or two representative cells** from each slide.

ONION IN DISTILLED WATER	ONION IN 20% SALINE SOLUTION

5. Add the following labels to your diagrams: **cell wall, nucleus, nucleolus, cytoplasm**.

6. In one of your slides, you should be able to see the onion **cell membrane**.

   On which slide is the cell membrane visible? _____

   Add a label for the cell membrane to your drawing.

7. If we took cells from a **leaf at the top of the onion plant** instead of from the onion bulb, what **additional organelle** would you expect to see in those cells?

   Why isn't the organelle you listed present in **your** onion-skin sample?

8. **What happened** to the onion cells that were soaked in the **20% saline solution**? **Support your answer** by mentioning details you observed under the microscope.

 Check your diagram labels and answers with your instructor before you continue.

☑ Comprehension Check

1. Assume that the interior of an onion cell is **98% water molecules**. The remaining **2%** is composed of various dissolved particles.

   The salt solution you applied to the slide is _____ % **water** molecules and _____ % **salt** particles.

2. Using the information in **Question 1, add an arrow** to the **picture** you drew of the **onion cells in saline solution**, showing the **direction of water movement**.

3. **(Circle one answer.)** The process that occurred to the onion cells in salt solution was an example of **diffusion/osmosis**.

   Explain your answer.

4. The onion cell membranes allowed water to enter and leave the cells freely, but **did not allow** salt particles to diffuse into the cells.

   This is an example of a _____ **permeable membrane**.

5. A student performed the onion cell experiment using **human blood cells**. When the student placed the blood cells in **salt water**, they shriveled up and died. When the student placed other blood cells in **distilled water**, they filled up, exploded, and died.

   What cellular organelle present in onion cells, but **not** blood cells, prevented the leaf cells from shriveling or exploding during your experiment? _____

Self Test

1. Of the cell organelles studied in this exercise, **list three** that are found in plant cells but not in animal cells.

2. In **Figure 4-6, draw arrows** showing the movement of the water molecules.

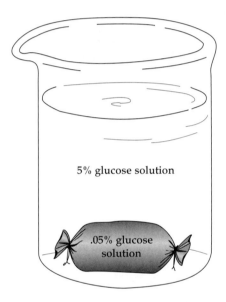

5% glucose solution

.05% glucose solution

FIGURE 4-6. Movement of Water Molecules

3. A patient with partial kidney failure enters the hospital. The doctor tells the patient that the kidney is responsible for filtering and removing waste molecules from body fluids. The doctor recommends use of a **kidney dialysis** machine. Using your knowledge of **diffusion, osmosis, and dialysis**, what do you think the dialysis machine will do this patient?

4. You are admitted to the hospital with severe dehydration. A student nurse is directed to give you a transfusion of **blood plasma** to replace your lost body fluids. The nurse gives you a transfusion of **distilled water** by mistake. **Explain** what will happen to your **red blood cells**. In your explanation, use the following words: **cell membrane, osmosis, high concentration, and low concentration**.

5. Will a sugar cube dissolve faster in iced tea or hot tea? _____

    Explain your answer. In your explanation, use the following words: **diffusion, high concentration, low concentration, molecular movement, and collisions**.

6.   You are in a restaurant that is divided into "smoking" and "nonsmoking" sections. There is a partition between the two sections, but the partition does not reach the ceiling. Despite the fact that you are seated in the nonsmoking section, you are bothered by the smell of cigarette smoke during your meal. Explain why the cigarette smoke was present in the nonsmoking section of the restaurant. In your explanation, use the following terms: **diffusion, molecules, higher concentration, and lower concentration**.

Investigating Cellular Respiration

Objectives

After completing this exercise, you should be able to:

- explain the process of aerobic cellular respiration in plants and animals
- summarize experimental results that show that microorganisms also perform aerobic cellular respiration
- compare the rate of aerobic cellular respiration and carbon dioxide production while at rest with that during exercise
- explain the similarities and differences between aerobic respiration and anaerobic respiration and give examples
- explain the similarities and differences between ethanol fermentation and lactic acid fermentation
- apply your knowledge of aerobic and anaerobic respiration to real-life situations.

CONTENT FOCUS

Cellular respiration is the process by which living organisms convert the **chemical energy** in food (**organic molecules**) to a useful form for cellular functions. This conversion process is comparable to the production of electrical energy to run machines in your home. A television cannot take energy directly from coal, gasoline, or even nuclear fuel. These fuels must be converted to electricity by your community power plant before they can be used by your home appliances.

Cells must transfer the energy stored in food molecules into **adenosine triphosphate (ATP)**, a form of energy that can be used to run cell activities. If you read the ingredients on a candy wrapper or a box of corn flakes, however, you will not find ATP listed. The energy in the foods has to be **transferred** into ATP. This conversion process is called **aerobic cellular respiration**.

During **aerobic respiration**, living organisms extract energy from the chemical bonds in food molecules (such as carbohydrates, fats, or proteins) and convert that energy into ATP.

The process of cellular respiration can be summarized with this equation:

food molecules + oxygen → carbon dioxide + *ATP energy* + water + heat

Aerobic cellular respiration is not the same as breathing. It occurs continuously, day and night, in all living cells. If it stops, the cell will quickly die. Aerobic respiration is not a single chemical reaction, but involves as many as 50 intermediate steps and the formation of a number of different compounds before carbon dioxide and water are finally produced. Specific enzymes are needed to control each step. For all living organisms, the chemical reactions of cell respiration are amazingly similar. Many pathways are virtually identical, even between such dissimilar organisms as bacteria and human beings.

The process of aerobic cellular respiration **uses oxygen** and **produces carbon dioxide**. For this reason, cellular respiration cannot take place without gas exchange. Thus, when you breathe in, oxygen is carried by the bloodstream to all the cells of your body. As blood circulates through your tissues, it picks up the carbon dioxide produced by cellular respiration and transports it to the lungs to be removed. Every time you exhale, carbon dioxide is released.

To sum up, aerobic respiration is a process that uses oxygen to burn the food (fuel) we consume and produce useful ATP energy. The more energy your body needs, the more fuel and oxygen you must take in. It should now be clear why you increase your breathing rate (and therefore your oxygen intake) when you exercise. During exercise, you also increase the amount of carbon dioxide you produce, since you burn more glucose to produce ATP.

ACTIVITY 1

EVIDENCE THAT CO_2 IS RELEASED
DURING CELLULAR RESPIRATION

Preparation

From the following equation, you can see that organisms produce carbon dioxide gas (CO_2) during aerobic cellular respiration.

food molecules + oxygen → *carbon dioxide* + ATP energy + water + heat

Bromothymol blue is a solution that turns yellow when CO_2 is added and turns back to blue when CO_2 is removed. We will use bromothymol blue as an **indicator** that will tell us if CO_2 is produced.

1. Work in groups.

   In your classroom, you will find a demonstration of cellular respiration. The demonstration setup consists of a paper bag covering a rack of sealed test tubes containing **bromothymol blue and sample organisms (see Figure 5-1)**.

Note:

Read through the instructions COMPLETELY and record your hypotheses in Table 5-1 BEFORE LOOKING AT THE DEMONSTRATION!!

Demonstration Setup:

Tube 1	No sample
Tube 2	Seedling (young plant)
Tube 3	Plant seeds
Tube 4	Plant leaf
Tube 5	Cricket
Tube 6	Pebble

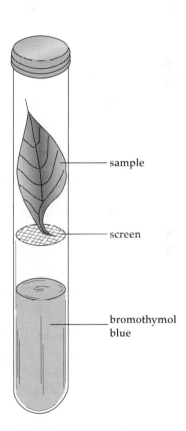

The samples are suspended on a piece of screen so that they will not fall into the bromothymol blue solution.

FIGURE 5-1. Appearance of Experimental Test Tubes

2. **Form a hypothesis for each test tube.** Which of the samples will produce CO_2 and change the color of the indicator solution from blue to yellow?

   Enter your hypotheses in the first column of **Table 5-1**.

TABLE 5-1

WHICH TUBES CONTAIN CARBON DIOXIDE?

SAMPLE	HYPOTHESIS WILL CO_2 BE PRODUCED? (YES OR NO)	RESULTS COLOR OF INDICATOR SOLUTION	CONCLUSION WAS CO_2 PRODUCED? (YES OR NO)
Tube 1: No sample			
Tube 2: Seedling (young plant)			
Tube 3: Plant seeds			
Tube 4: Plant leaf			
Tube 5: Cricket			
Tube 6: Pebble			

3. Carefully lift the paper bag and **observe** the samples. **Record your observations** in the **Results column** of **Table 5-1**.

☑ Comprehension Check

1. On the basis of your observations, which tubes showed evidence of CO_2 production? Record your answers in the **Conclusions column of Table 5-1**.

2. Was there a control tube in this experiment? _____

    If so, which tube was the control? _____ **Explain your answer**.

3. **On the basis of your results**, which of the sample organisms was performing cellular respiration? **Explain your answer.**

4. **(Circle one answer.)** If one test tube contained a **dead cricket**, the amount of CO_2 in the test tube would be **more/less/exactly** the same compared to the tube with the living cricket.

5. **(Circle one answer.)** Seeds contain plant embryos. On the basis of the results of this experiment, the embryos in the tested seeds are **active/dormant (inactive)**.

Check your answers with your instructor before you continue.

ACTIVITY 2

DO MICROORGANISMS PERFORM CELLULAR RESPIRATION?

Preparation

Yeasts are microscopic fungi that are commercially very important. They perform aerobic cellular respiration using pathways similar to those found in larger plants and animals.

Methylene blue dye can be used as an **indicator** for cellular respiration in yeast.

Aerobic respiration releases hydrogen ions and electrons that are picked up by the methylene blue dye, gradually turning the dye **colorless**. The mitochondria of yeast cells undergoing cell respiration will appear as a clear area surrounded by a ring of light blue cytoplasm. The nucleus may be visible as a small, darkly stained spot.

If cellular respiration is not taking place, the mitochondria will absorb the blue dye and will not turn colorless. The cells will appear to have a large, darkly stained central area surrounded by the ring of light blue cytoplasm.

1.   Work in groups. Get the following supplies: **slides, coverslips, yeast suspension, a pipette, and a dropper bottle of methylene blue dye**.

2.   Place a **drop of yeast suspension** on a clean microscope slide. Add **one small** drop of **methylene blue** dye and place a **coverslip** over the mixture. Observe the yeast cells with the **high-power objective**.

3.   Is cell respiration occurring in any yeast cells on your slide? _____

      What percentage of the yeast cells is **not** undergoing cell respiration? _____ (Make an estimate.)

      Why **isn't** cell respiration taking place in all cells?

4.   **Draw** one yeast cell that is undergoing cell respiration and one that **isn't**. Make the drawings **large and clear**. Label the **cytoplasm, nucleus (if visible), and mito-chondria** of your yeast cells.

YEAST CELL CELL RESPIRATION OCCURRING	YEAST CELL CELL RESPIRATION ABSENT

Check your drawings with your instructor before you continue.

ACTIVITY 3

COMPARISON OF CLASSROOM AIR
WITH EXHALED AIR

Preparation

This experiment uses **limewater** (a very concentrated **calcium carbonate solution**) as an **indicator** solution. Limewater turns cloudy when CO_2 is added to it.

1. Work in groups. Get the following supplies: **two small beakers, two drinking straws, a rubber bulb, and a container of limewater**.

2. **Fill** each beaker **about half full with limewater**. Label the beakers 1 and 2.

> ### Note:
>
> **Read through the instructions COMPLETELY before continuing.**

3. Do you think bubbling **room air into Beaker 1** will cause the limewater to turn cloudy? **Enter your hypothesis in Table 5-2**.

4. Do you think bubbling **your exhaled air into Beaker 2** will cause the limewater to turn cloudy? **Enter your hypothesis in Table 5-2**.

TABLE 5-2		
EFFECT OF BUBBLED AIR ON LIME WATER		
	HYPOTHESIS (YES OR NO)	RESULTS
Will room air cause the lime water in Beaker 1 to turn cloudy?		
Will exhaled air cause the lime water in Beaker 2 to turn cloudy?		

5. Attach the rubber bulb to one of the drinking straws. Using the drinking straw with the bulb, continuously bubble air into Beaker 1 for one **minute**.

   Observe the beaker during the bubbling process and **record** your results in Table 5-2.

6. Using another drinking straw, **very gently** bubble your exhaled air into **Beaker 2** for **one minute**.

Caution!

Be careful not to suck the solution into your mouth!

   Observe the beaker during the bubbling process and record your results in **Table 5-2**.

7. When you have completed your experiment, empty your beakers into the **waste container**.

✔ Comprehension Check

   1. In one or two sentences, **summarize the results** of this experiment.

   2. On the basis of your results, **what gas was bubbled** into Beaker 2? _____

   3. In reference to **Question 2**, what process **produced** this gas?

   Check your answers with your instructor before you continue.

ACTIVITY 4

EFFECT OF EXERCISE ON CARBON
DIOXIDE PRODUCTION

Preparation

Breathing is controlled by a "breathing center" in the medulla of the brain. This center is activated by a **buildup of carbon dioxide in the blood and/or a low oxygen concentration**, such as occurs at high elevations. This experiment will compare the **amount** of CO_2 produced during exercise with that produced at rest.

We will use **bromothymol blue** as an **indicator** that will tell us **how** much CO_2 is produced. Recall that bromothymol blue turns **yellow** when CO_2 is **added** and returns to **blue** when CO_2 is **removed**.

1. Work in groups.

    Get the following supplies: **three 400-ml beakers, bromothymol blue solution, ammonia solution, a plastic bag, a coupler tube, a mouthpiece, a rubber band, and an air stone assembly**.

2. Fill each beaker **with 200 ml of bromothymol blue indicator solution. Label** the beakers **1, 2,** and **3**.

3. **Attach the coupler tube** to the plastic bag and hold the tube in place with the rubber bands (as shown in **Figure 5-2**).

Testing Room Air

1. Fill the plastic bag with **room air** by pulling it through the air ("rounding out" the bag).

    Quickly plug the bag with the **air stone assembly. Be careful not to lose the air**.

2. Place the **air stone** into the indicator solution in **Beaker 1** (see **Figure 5-2**).

    Squeeze the air out of the plastic bag so that it bubbles into the beaker.

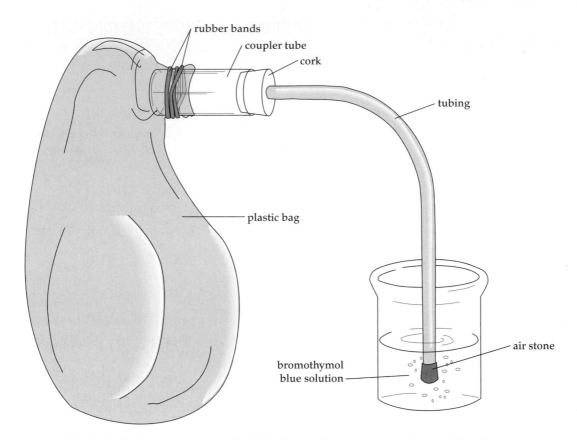

FIGURE 5-2. Apparatus to Bubble Captured Air into Bromothymol Blue Solution

Ammonia Treatment

1.   Using the dropper bottle, add **ammonia** to **the beaker one drop at a time** and **gently** stir after each drop.

     Be sure to **keep an accurate count** of the number of drops added.

2.   Continue adding ammonia until the indicator solution changes to a **light green color and remains green for 30 seconds**.

3.   **Record** the number of drops you added in **Table 5-3**.

     The amount of ammonia needed to change the indicator solution from **blue to green** will be used as an estimate of **the amount of CO_2 present** in the room air.

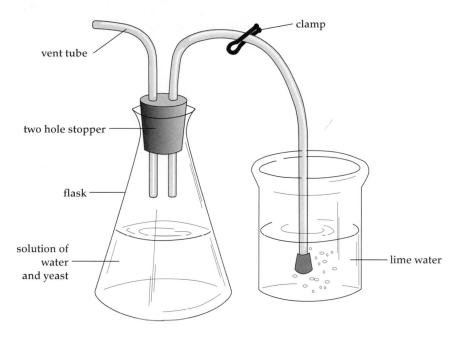

FIGURE 5-4. Setup for Fermentation Experiment

3. **Wearing eye protection,** gently remove the stopper from **Flask B and add one tablespoon of sucrose.**

   Replace the stopper. Record the time when you added the sucrose._____

4. **Immediately** after adding the sucrose, remove the clamp from the air stone tube and **switch it to the vent tube. Switch the clamps on Flask 1 and Flask 2.**

5. Check on your experiment every **five minutes** for the next **15 minutes. Record** the results for **each five-minute period** in Table 5-4.

TABLE 5-4 OBSERVATIONS OF YEAST FLASKS						
OBSERVATIONS	EXPERIMENTAL FLASK A (WATER AND YEAST)			EXPERIMENTAL FLASK B (WATER, YEAST, AND SUCROSE)		
	5 MIN.	10 MIN.	15 MIN.	5 MIN.	10 MIN.	15 MIN.
Color of liquid in yeast flask						
Color of lime water in beaker						
Presence of bubbles in yeast flask						

6. Did the results for Flasks A and B differ? Explain your answer by mentioning facts collected in this experiment.

7. What gas formed the bubbles observed in the flask(s)?_____

    Explain your answer.

Caution!
Eye protection is required for the following procedures!

8. **Wearing eye protection**, carefully remove the stopper from **Flask A** and sniff the released air. **Repeat** the process with **Flask B. Describe** the results.

✓ Comprehension Check

1. A yeast cell undergoing fermentation produced **six ATP molecules**. How many glucose molecules were used?_____ **Explain your answer**.

2. If the same cell was undergoing aerobic cellular respiration instead of fermentation, but using the same number of glucose molecules, how many ATP molecules would be produced?_____ **Explain your answer**.

Check your answers with your instructor before you continue.

Note:
Lactic acid fermentation can only satisfy the body's short-term energy needs. If the oxygen shortage continues too long, lactic acid accumulates in your muscle tissue, lowering the pH. Eventually, this acidic environment inhibits the activity of enzymes that are needed for muscle function. Muscle fibers lose their ability to contract, a process called muscle fatigue.
The accumulated lactic acid can damage muscle proteins, producing the soreness and pain often associated with lengthy, strenuous exercise.
When oxygen is available again, the lactic acid is converted back to pyruvic acid and aerobic respiration can continue as before.

Self Test

1.   A rock covered with green spots is placed on a screen in a test tube above some bromothymol blue solution; after several hours, the solution remains blue. What can you **conclude** from this experiment?

2.   You and several other students observe that dead animals along the roadside often increase in size several days after they have been killed. In the spirit of scientific investigation, you collect the gas from inside one of these dead animals and bubble it through limewater. The limewater turns cloudy.

     a. What does the cloudy result indicate?

     b. If the animal is dead, how are these results possible? **Explain your answer in detail**.

3.   Which produces the most energy in the form of ATP?

     a. Aerobic respiration
     b. Anaerobic respiration
     c. Alcohol fermentation
     d. All of the above produce the same amount of ATP energy but through different chemical reactions
     e. None of the above produce ATP

4. Carbon dioxide is passed into a solution of bromothymol blue indicator until the solution turns yellow. A sprig of elodea is then placed into this yellow solution. After a few hours in the sunlight, the yellow solution turns blue again. **Suggest an explanation** for the color changes observed in the bromothymol blue.

For Questions 5 through 8, use the following three answers. An answer may be used once, more than once, or not at all. For each question, **explain your answer**.

a. Process occurs only under aerobic conditions
b. Process occurs only under anaerobic conditions
c. Process occurs under both aerobic and anaerobic conditions

5. _____ Manufacturing of beer or wine

    Explanation:

6. _____ ATP production

    Explanation:

7. _____ Accumulation of lactic acid in a runner's muscles causes her to drop out of the race

    Explanation:

8. _____ Production of 36 ATPs, water, and carbon dioxide from one glucose molecule

   Explanation:

9. If an elephant's cells were to lose their ability to perform aerobic cellular respiration, what would most likely happen?
   a. The cells would switch over to anaerobic respiration and the elephant would be fine
   b. The cells would switch over to lactic acid fermentation and the elephant would survive but would have achy muscles
   c. The cells would switch over to alcohol fermentation and the elephant would survive in a drunken stupor
   d. The cells would die because the elephant's energy needs could not be met using anaerobic metabolic pathways
   e. The cells could live on stored energy for an extended period and cellular respiration would not be necessary for several years

   Explain your answer.

Nutrient Analysis
of Foods

Objectives

After completing this exercise, you should be able to:

- test an unknown sample for the presence of sugars, starch, protein, and lipids using indicator chemicals
- relate the nutrient content of a food to its original function in plants and animals
- discuss the benefits and drawbacks of using carbohydrates versus lipids as energy storage molecules for embryos
- use the Federal Dietary Guidelines to analyze the nutritional value of your meals.

CONTENT FOCUS

All living organisms are composed of various types of **organic molecules** such as carbohydrates, lipids, proteins, and nucleic acids that make up their body tissues. Each of these four types of organic molecules (carbon-based compounds) has different functions in the body. The foods we eat are obtained from plants and animals. These foods are needed not only as a source of energy but also because they provide many essential nutrients. Just as different parts of your body have different structures to support their functions, the same is true of other animals and plants. The **types** of **organic chemicals present** in different body tissues are related to the **function** of those tissues.

In this exercise, you will be analyzing various food samples to determine which organic compounds are present. Carbohydrates, lipids, proteins, and nucleic acids **cannot be seen directly. Indicator chemicals** can show you if these organic compounds are present in a food sample by **changing color**. Each type of indicator chemical is specific for one type of organic compound.

Analyzing foods for organic compounds using indicator chemicals can present some problems. Often, the **food itself has colors** that may interfere with the test results. It is easier to determine if an appropriate color change has occurred if you have a **standard for comparison**. Therefore, we must first establish a set of **positive and negative standards (controls)** for using them for comparison in later experiments.

ACTIVITY 1   POSITIVE AND NEGATIVE INDICATOR TESTS

1.   Work in groups. Get the following supplies: **a ruler, a marking pencil, six test tubes, a test rack, a test tube holder, and a large beaker**.

2.   Using the ruler, place a **line** on each test tube **1 cm from the bottom** of the tube. **Number each test tube 1 through 6**. Place the marked test tubes in the test tube rack.

Benedict's Test for Simple Sugars

1.   The chemical indicator in Benedict's solution reacts with monosaccharides and some disaccharides, but **only when the indicator solution is heated**.

     Place the large beaker, half filled with tap water, on a hot plate. Set the temperature control to high until the water boils.

2.   Fill **test tube #1** to the line with **glucose** solution. Fill **test tube #2** to the line with **distilled water**.

3.   Fill the dropper with Benedict's solution. What color is Benedict's solution?

     _____

4.   To **test tubes #1 and #2**, add **one dropper full of Benedict's solution**.

     Shake the test tubes to mix.

5.   **Using a test tube holder, carefully** place the tubes into the boiling water for 2 to 5 minutes.

Caution!
Hot glass looks exactly like cold glass! Use the test tube holder to remove test tubes from the boiling water!
Don't leave the test tube holder clamped on the test tube while in the boiling water. Hot metal looks exactly like cold metal!

6.   **Using the test tube holder** remove the tubes and **observe** the color in each test tube.

If you see a color change ranging from **green through yellow, orange, or red**, this is a positive test for simple sugars.

If the color in the tube has **not changed**, this is a **negative** test for simple sugars.

Note:
Save tubes #1 and #2 for use in later experiments.

---

7.   What happened to the **color** in **tube #1**?_____

8.   What happened to the **color** in **tube #2**?_____

✓ Comprehension Check

1.   Which tube showed a positive test for glucose?_____

2.   On the basis of the **results of Benedict's test, glucose** could be correctly identified as belonging to which of the following groups?

   a. Lipids   d. Simple sugars
   b. Starches   e. Both b and d are correct
   c. Proteins

Iodine Test for Starch

1.   Fill **test tube #3** to the line with **starch** solution. Fill **test tube #4** to the line with **distilled water**.

2.   Fill the **dropper with iodine**. What color is iodine?_____

3.   To **both tubes**, add one **dropper full of iodine. Shake** the test tubes to mix.

   Observe the color in each test tube.

   What color is the liquid in **test tube #3**?_____

   What color is the liquid in **test tube #4**?_____

4.   Which tube showed a positive test for starch?_____

5.   Place **one drop of iodine** in the box.

On the basis of the results of the iodine test, what can you **conclude about the paper** in the laboratory manual?

Biuret Test for Protein

1.   Fill **test tube #5** to the line with **albumin (egg white)** solution. Fill **test tube #6** to the line with distilled **water**.

2.   Fill the dropper with **Biuret solution**. What color is Biuret solution?_____

3.   To **tubes #5 and #6**, add one dropper full of **Biuret solution. Shake** the tubes to mix.

4.   **Observe** the color in each test tube.

5. What happened to the color in **tube #5?**

6. What happened to the color in **tube #6?**

7. Which tube showed a **positive** test for **protein?**_____

Sudan IV Test for Lipids

1. Get the following supplies: **one piece of filter paper, forceps, a petri dish, and dropper bottles of vegetable oil, distilled water, and Sudan IV stain.**

2. Place the filter paper on a **clean** piece of paper (**not** on the laboratory counter, which may be dirty).

   Using a **pencil**, draw **two** circles **the size of a dime**, spaced **apart** on the filter paper. Label the first circle **"oil"** and the second circle "**water."**

3. Place **one drop of oil** into the circle labeled "oil" and **one drop of distilled water** into the circle labeled **"water."**

4. **Set the filter paper aside to dry**. If needed, use a hair dryer **on low power** to evaporate excess liquid.

5. When the filter paper is **completely dry**, place it in the **petri dish** and cover the paper with **Sudan IV solution**. Let the paper soak in the stain for **three minutes.**

6. While the paper is soaking, get a **glass bowl** from the **supply area** and **fill it** with **distilled water.**

7. When the three minutes are up, place the filter paper into the bowl of distilled water. **Rinse gently** for **one minute.**

   Remove the filter paper from the water with the forceps and **observe the color of the two circles**.

8. **A dark red spot** indicates a **positive** test for lipids.

    A **pale pink** color, **no different from the rest of the paper**, should be considered **negative**.

9. **Record** the results of your test (positive or negative) below. Enter a "+" if the test was **positive** and a "−" if the test was **negative**.

    "oil" circle _____ "water" circle _____

> ### Note:
> Save your test paper for use in later experiments.

✔ Comprehension Check

Using the results from the tests you have just completed, fill in **Table 6-1**.

TABLE 6-1 INDICATOR TESTS			
INDICATOR TEST	TESTS FOR	NEGATIVE RESULT (COLOR)	POSITIVE RESULT (COLOR)
			green through red
	starch		
Sudan IV Test			
Biuret Test		light blue	

Check your answers with your instructor before you continue.

ACTIVITY 2   TESTING FOOD SAMPLES

Preparation

Have you ever wondered what is really in the foods you choose? Is a food high in fat or protein? Is it a good source of carbohydrates? The answers to these questions become clearer when food samples are analyzed to determine which organic compounds are present.

In the next activity, you will analyze several commonly eaten foods and determine their nutrient content. Solid food samples have been blended into liquid, for easier testing.

1. For each of the following foods, form a **hypothesis** about which organic compounds it will contain.

    Record your hypotheses in **Table 6-2**.

For each hypothesis, enter a "+" if you think the test will be **positive** and a "−" if you think the test will be **negative**.

TABLE 6-2 **HYPOTHESES FOR FOOD EXPERIMENTS**				
FOOD	SIMPLE SUGARS	STARCH	PROTEIN	LIPID
Peanut butter				
Tuna				
Refried beans				
Milk				
Hamburger				
Lettuce				

2. Work in groups. Your instructor will assign several foods for your group to analyze.

    Your group will **share its data** with other groups that are analyzing different foods.

3. For each food sample to be analyzed, get **three clean test tubes**. You will also need **one piece of filter paper**.

4. Following the **test procedures exactly as you did in Activity 1**, analyze each food sample for the presence of **simple sugars, starch, protein, and lipids**.

5. **Compare** your test results to the **controls** you **saved from Activity 1**.

6. **Record** the results of your tests (positive or negative) in **Table 6-3** and also on the **master chart at the front of the room**.

   Enter a " **+** " if the test was **positive** and a " **−** " if the test was **negative**.

7. **Complete Table 6-3** by entering the results from the **master chart** at the front of the room.

TABLE 6-3 RESULTS OF FOOD EXPERIMENTS				
FOOD	SIMPLE SUGARS	STARCH	PROTEIN	LIPID
Peanut butter				
Tuna				
Refried beans				
Milk				
Hamburger				
Lettuce				

◢ Comprehension Check

1. List the foods tested that contain **all three** of the following: **carbohydrates, proteins, and lipids**.

It is possible to relate the nutrient content of a food to its original function in plants and animals. Using the information provided below, **explain** why certain nutrients occur in high levels in particular foods.

2. This food comes from the leaf of a plant. Leaf cells perform photosynthesis.

   Which of the six foods is being described?_____

   (Circle one answer.) If photosynthesis is taking place, I would predict a positive indicator test for **simple sugar/lipid/protein**.

3. Of the foods tested, which had high levels of **lipids?**_____

4. Of the foods tested, which use **lipids** as a high energy source for young animals?_____

5. **Muscle tissue** is composed of **interwoven protein fibers**. Using this information, **name two foods** made of **muscle** tissue that gave you a **positive** test for **protein**.

6. The hamburger and tuna tested **negative** for starch. **Explain** why foods of this type would **not** have starch present.

Check your answers with your instructor before you continue.

ACTIVITY 3   AN EMBRYO IN A PEANUT SEED

Preparation

A seed contains a **plant embryo** packaged with its **food supply**. The embryo and stored food are protected by a tough seed coat. A developing embryo has the beginnings of a stem, tiny leaves, and a root. You can see the tiny stem, leaf, and root structures of an **embryonic peanut plant** if you carefully pull the two peanut halves apart.

The nutrients in seeds are not only used by germinating plant embryos but also consumed by animals, including humans.

1. In your classroom, you will find a demonstration of a **separated peanut seed** that can be viewed with a dissecting microscope.

2. Look for the following structures and label them on the diagram in **Figure 6-1:** **root, stem, leaves,** and **stored food supply**.

FIGURE 6-1. Peanut Seed

✓ Comprehension Check

1. Food may be stored for the use of an adult plant or as energy for the development of an embryo. The stored food is packaged with an embryo into a structure called a **seed**.

   List **three commonly eaten foods** produced by plants or animals to provide stored energy for embryos:

   _____   _____   _____

2. Of the foods tested in Activity 2, which use **lipids** as an energy source for a **developing embryo?**

3. **(Circle all correct answers.)** The following probably have high lipid levels:

   a. Chicken egg   c. Sunflower seed   e. Coconut

   b. Carrot   d. Spinach   f. Almond

   Explain your answer.

4. Considering the information gained from your **food analysis experiments**, what is one possible explanation for a small seed sprouting more quickly than a larger seed?

Check your answers with your instructor before you continue.

ACTIVITY 4   DIETARY INTAKE AND GOOD HEALTH

Preparation

When you eat a meal, you are probably consuming a combination of carbohydrates, proteins, and lipids. Imagine this situation. You've been stuck in classes all morning (just grabbed a bag of peanuts from the candy machine), worked in the afternoon, picked up the baby-sitter, and now you're late for a meeting of your study group. Of course, you're starving! On the way to the library, you stop off at your favorite fast food restaurant to refuel. You order a double hamburger with cheese, a large order of fries, and a chocolate milk shake.

One person in your study group is taking a nutrition course and has been studying dietary analysis. He was shocked to discover that his normal intake was not even close to the guidelines for a healthy diet. As he tells you about this, you wonder whether your diet is any better. Below, you will find the instructions provided by your friend for calculating the percentage of carbohydrates, protein, and fat in your diet. He also thoughtfully provided you with a copy of the government recommendations for a healthy diet (**Table 6-4**). **Table 6-5** summarizes your meals.

TABLE 6-4		
RECOMMENDED DIETARY GUIDELINES		
NUTRIENT	RECOMMENDED PERCENTAGE OF DAILY CALORIC INTAKE	
Carbohydrates	55–60%	
Protein	10–15%	
Fat	no more than 30%	
1 g carbohydrate contains 4 kcal	1 g protein contains 4 kcal	1 g fat contains 9 kcal

Follow the instructions below to calculate the percentage of carbohydrate, protein, and fat you consumed today.

1.   Begin your calculations with the hamburger. **Multiply** the grams of carbohydrates in the hamburger by **4** (since there are 4 kcal per gram of carbohydrate).

      Record the answer in Table 6-5. Perform the same calculations for the **protein and fat** in the hamburger (using the appropriate number of kilocalories) and record your answers.

2.   Calculate the **kilocalories** for the other foods listed in **Table 6-5** and record your answers.

3.   Total each column in **Table 6-5** and **record** your answers.

Add the total **kilocalories** from carbohydrates, **protein, and fat** and record the answer at the bottom of **Table 6-5**.

T A B L E   6 - 5
DIETARY CALCULATIONS

Food Eaten	Carbohydrate (g)	Carbohydrate (kcal)	Protein (g)	Protein (kcal)	Fat (g)	Fat (kcal)
Double beef hamburger with cheese	54		51		60	
Fries (large order)	43		5		20	
Chocolate milk shake	49		9		10	
Bag of peanuts (2 ounces)	10		14		28	
Column Totals						
Total kcal:						

4.   Calculate the **percent of the day's diet that came from carbohydrates, protein, and fat**.

_____ % of carbohydrates in diet $=$ $\dfrac{\text{total kcal from carbohydrates}}{\text{total kcal consumed}} \times 100$ $=$ %

_____ % of protein in diet $=$ $\dfrac{\text{total kcal from protein}}{\text{total kcal consumed}} \times 100$ $=$ %

_____ % of fat in diet $=$ $\dfrac{\text{total kcal from fat}}{\text{total kcal consumed}} \times 100$ $=$ %

_____
                                                                                                      100%

5.   Compare your calculated percentages to the federal government's recommended values.

Was your intake today in line with recommended dietary goals?_____

If not, which nutrients did not match the guidelines?

6.   Place an X in front of any of the following that are good suggestions for improving your day's food intake.

    ___ Bring an apple or an orange for a snack instead of the peanuts.

    ___ Replace the fries with a baked potato.

    ___ Replace the hamburger with pizza.

    ___ Replace the milk shake with low-fat milk or juice.

    ___ Eat a candy bar instead of the peanuts.

7.   If your body requires **1500 kcal per day**, how many excess kilocalories did you eat in the day described in **Table 6-5**? _____

If you consumed the same amount of excess kilocalories each day for the last three weeks, **how many excess kilocalories** did you accumulate? _____

Every time you accumulate 3500 excess kilocalories, you gain a pound.

How many pounds have you gained over the last three weeks? _____

8.   You want to lose this excess weight, but with your schedule, you're too busy to exercise regularly. However, if you park in the furthest lot from the building, you can get to class in 15 minutes of quick walking. Every time you do this, you will burn **150 kcal**.

If you walk quickly to and from the parking lot once a day, you can lose the weight you gained in only _____ days.

Check your answers with your instructor before you continue.

Self Test

1. You add **Biuret solution** to your morning orange juice. The Biuret solution does not change color. What can you **conclude** from this experiment?

2. What do you think would happen if you placed a drop of iodine on your baked potato at dinner?

    On your steak?

    Explain your answer.

3. Match the foods in Table 6-6 with their functions in the living organism they come from. For example, why does a cow produce milk? (Hint: Not for your morning cereal!)

TABLE 6-6	
MATCHING FOODS AND THEIR FUNCTIONS	
FOOD	FUNCTION IN PLANT OR ANIMAL
Peanut butter	
Tuna	
Bean	
Milk	
Hamburger meat	
Lettuce	

Factors That Affect Enzyme Activity

Objectives

After completing this exercise, you should be able to:

- discuss the basics of enzyme function in cells
- explain the relationship between the three-dimensional structure of proteins and enzyme function
- describe the effects of various environmental conditions on protein denaturation
- explain the activity of digestive enzymes in food vacuoles.

CONTENT FOCUS

Within living systems, chemical reactions require specific **enzymes** to assist and speed the rate of the reactions (these "helper" molecules are known as **catalysts**). Enzymes are **proteins,** and they are very specific, each working with only one or a very few chemical compounds.

In addition, different enzymes work best under different environmental conditions.

In order for a protein to function correctly, it must be folded into a specific three-dimensional shape. In enzyme molecules, folding creates a region called the **active site,** where molecules can bind to the enzyme and a reaction can take place.

If environmental factors (such as pH, temperature, or salinity) lead to changes in the specific three-dimensional shape of the enzyme's active site, the enzyme may not function correctly. When an enzyme's folding is altered, the enzyme is said to be **denatured.**

The following activities will demonstrate how enzymes work and the effect of some environmental conditions on protein structure.

ACTIVITY 1   DEMONSTRATION OF ENZYME ACTIVITY

Preparation

Amylase is an enzyme that hydrolyzes starch (a polysaccharide). In humans, amylase is present in the mouth and the small intestine.

To demonstrate the action of amylase on starch molecules, two indicator tests will be used.

Iodine is a chemical indicator solution. It changes color (turns dark blue or black) in the presence of **starch**. In the presence of monosaccharides or disaccharides, iodine retains its normal light brown color.

Benedict's solution is a chemical indicator for the presence of **simple sugars** (monosaccharides and some disaccharides). In the presence of simple sugars, Benedict's solution changes color from turquoise blue to one of the following colors: green, yellow, orange, or red. Green represents the smallest amount of simple sugar and red the highest. Benedict's test differs from the iodine test in that the reaction only takes place when the solution is heated.

1. Work in groups. Get the following supplies: **four large test tubes, two glass stir-ring rods, a test-tube rack, a test-tube holder, two graduated 10-ml pipettes with manual dispenser, a hot plate, a dropper bottle of iodine solution, a dropper bottle of Benedict's solution, a stock bottle of starch solution, a stock bottle of amylase solution, and a large beaker.**

2. Place the large **beaker**, about **one-third filled** with tap water, onto the **hot plate**.

    Set the temperature control to high until the water boils.

3. **Label the test tubes as follows: B-1, B-2, I-1 and I-2.**

    Make sure the bottle of starch solution is **well mixed**.

    Using a **10-ml graduated pipette**, transfer **15 ml of starch solution** from the stock container into **each** of the large test tubes.

4. To the test tube labeled **B-1**, add **two droppers of Benedict's solution**.

    What color is the liquid in **tube B-1**? _____

5. The chemical indicator in Benedict's solution **will only react with simple sugars when it is heated up**.

    Using a **test-tube holder, carefully** place the **tube B-1** into the boiling water for **five minutes**.

Caution!
Hot glass looks exactly like cold glass! Use the test-tube holder to remove test tubes from the boiling water!
Don't leave the test-tube holder clamped to the test tube while in the boiling water. Hot metal looks exactly like cold metal!

6. **Using the test-tube holder, remove** the test tube from the boiling water and **observe** the color of the Benedict's solution.

    Do not dispose of tube B-1.

    Was your Benedict's test result for tube B-1 **positive or negative** ($+/-$)? _____

7. To the test tube labeled **I-1**, add two droppers of iodine solution.

    Do **not** add indicator solution to test tube **I-2**.

8. Observe the color of the indicator solution in **tube I-1**.

    Was your iodine test result **positive or negative** ($+/-$)? _____

    Do not dispose of tube I-1.

9. Using a **clean 10-ml graduated pipette,** add **15 ml of amylase solution to tubes B-2 and I-2**.

    Record the time when you add the amylase to the starch solution: _____

    Place a **glass stirring rod** into each tube: **B-2** and **I-2**.

    Set tubes B-2 and I-2 aside for **30 minutes**.

    Every five minutes, stir the contents of each tube with the rod.

10. After the 30-minute interval, add **two droppers full of Benedict's solution to tube B-2**.

    Add two droppers of iodine solution to tube I-2.

    Shake the test tubes to mix the contents.

11. Observe the color of the indicator solution in **tube I-2**.

    Was your iodine test result **positive or negative** ($+/-$)? _____

12. If the **iodine test is negative in tube I-2**, what would this tell you about the digestion of starch by amylase?

13. Imagine that you get a **positive result for the iodine test in tube I-2**. You form the **following hypotheses** about the reason for the positive test result:

    The amylase enzyme was inactive and therefore did not digest any starch.

    Suggest a method by which this hypotheses could be tested.

 Check your answers with your instructor before you continue.

14. **Using the test-tube holder**, place **tube B-2** in the boiling water for five minutes.

    Remove the test tube from the boiling water and **observe** the color of the Benedict's solution.

    Was your Benedict's test result **positive or negative (+ / −)?** _____

15. Do the test results **support your hypotheses? Explain your answer**.

✓ Comprehension Check

1. If your Benedict's test result was positive, where did the sugar come from?
   Explain your answer.

2. What was the purpose of the indicator tests conducted on tubes I-1 and B-1?

3. Why were tubes I-1 and B-1 saved until the end of the experiment?

Check your answers with your instructor before you continue.

ACTIVITY 2   ENZYME ACTIVITY IN FOOD VACUOLES

Preparation

Paramecium is a small one-celled organism found in freshwater. By means of rapid swimming motions, it funnels water containing bits of organic matter, such as yeast and bacteria, into a groove on its surface. From this groove, the materials enter the *Paramecium* through a "mouth" and are taken up by small organelles called **food vacuoles**.

The organic matter constitutes the *Paramecium's* food. Enzymes are released into the vacuoles and the food within them is digested.

1. Work in groups. Get the following supplies: **slides, cover slips, a compound microscope, a bottle of methyl cellulose (the bottle may also be labeled ®Protoslo), toothpicks, a *Paramecium* culture, and a solution of yeast stained with Congo Red.**

2. Make a wet mount using the *Paramecium* culture and add **one drop** of **red-stained yeast**.

    Add a **small** drop of the methyl cellulose to your slide, **gently** stir with toothpick, and add a cover slip.

3. Using the **scanning lens** of the microscope, locate one or more organisms.

    Switch to a higher power and locate the food vacuoles within the cell (refer to **Figure 7-1**).

    What is the initial color of the food vacuoles? _____

    Observe the food vacuoles in various individuals every few minutes for **fifteen minutes**.

    Describe the **color changes** that occur within the vacuoles.

4. Congo red is an **indicator chemical** that turns **blue** under **acidic** conditions and **red** under **basic** conditions.

    What does this tell you about the pH inside a food vacuole?

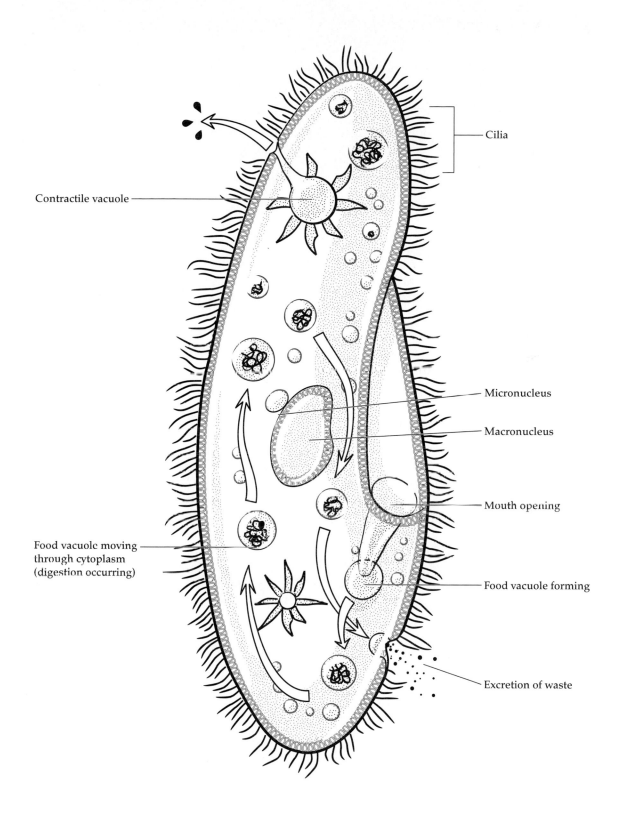

FIGURE 7-1. *Paramecium* Food Vacuole Formation

✓ Comprehension Check

1. **Pepsin** is a human digestive enzyme found in the stomach. If pepsin were present in *Paramecium*, do you think this enzyme could be active within a food vacuole? **Explain your answer**.

2. If a **base** (**alkaline solution**) was added to the yeast suspension before the *Paramecia* were fed, how might this affect the **color change** within the food vacuoles? **Explain your answer**.

3. If you ground up a "Tums" tablet to a fine powder and added it to the yeast suspension before the *Paramecia* were fed, how might this affect the **efficiency of digestion within the food vacuoles**?

Check your answers with your instructor before you continue.

ACTIVITY 3

EFFECT OF ENVIRONMENTAL CONDITIONS ON PROTEIN STRUCTURE

Preparation

During this laboratory exercise, you will be examining the effect of various environmental factors (temperature, salinity, and pH) on the folded structure of proteins. You will use chicken eggs, since they are among the few cells that are large enough to be observed without special equipment.

1. Work in groups. Get the following supplies: **five chicken eggs, five bowls, and stock bottles of the following solutions: pH 11 solution, pH 3, and 25% saline solution, a large beaker, clear plastic wrap, and a hot plate.**

2. Put about **400 ml of tap water** into the large beaker and set the hot plate on **high**. Heat the water to boiling.

3. **Form hypotheses** about the effect of each set of environmental conditions (**pH 11, pH 3, 25% saline, boiling, and room temperature**) on the yolk and the white of an egg.

    For example: no change will occur in the yolk or the white, the proteins in the egg yolk will denature, the proteins in the egg white will denature, the egg white will denature but not the yolk, etc.

    Record your hypotheses in **Table 7-1**.

TABLE 7-1		
RESULTS OF EGG EXPERIMENTS		
ENVIRONMENTAL CONDITIONS	HYPOTHESES	OBSERVATIONS OF THE EGG WHITE AND YOLK
pH 11		
pH 3		
25% saline solution		
Boiling water		
Room temperature		

4. Label **three** of the bowls as follows: **pH 11, pH 3, and 25% saline**.

    Carefully, without breaking the yolk, crack an egg into each of the three labeled bowls.

Caution!

Be careful not to spill the pH solutions on your skin. Rinse your hands or eyes thoroughly with water if contact occurs.

---

5. To the bowl labeled **pH 11**, add **100 ml of pH 11** solution.

    To the bowl labeled **pH 3**, add **100 ml of pH 3** solution.

    To the bowl labeled **25% saline**, add **100 ml of 25% saline** solution.

    Set the bowls aside undisturbed for **30 minutes**.

6. **Carefully, without breaking the yolk**, crack an egg into each of the two remaining bowls.

    To minimize odors, loosely cover the **pH 11** bowl with a sheet of clear plastic wrap.

    Set one of the bowls aside undisturbed for **30 minutes**.

7. Gently tip the egg out of the other bowl and into the beaker of boiling water.

    Let the water return to a full boil and boil the egg for **five minutes**.

    Turn the hot plate off and let the water cool to room temperature.

8. At the end of the 30-minute period, observe and record your observations for the yolk and white of each egg in **Table 7-1**.

    Do not dispose of any of the eggs until you have completed answering the remaining questions.

9. Did any of the environmental conditions damage (**denature**) the proteins in the egg? **Explain your answer**.

10. Test the egg white of the "room temperature" egg with pH paper.

   What is the approximate pH of the egg white? _____

   Is there a relationship between the pH results from the room temperature egg and your observations of the eggs exposed to **pH 3 and pH 11**? **Explain your answer**.

11. Was there a **control** in this experiment? If so, what was the control?

12. Why is a control desirable in an experiment of this type?

✔ Comprehension Check

1. Stomach conditions are quite acidic. Do cells of the stomach lining need protection from these acidic conditions? Why or why not?

2. Acid rain in the Northeastern United States can have a pH as low as 4.5 in some areas. How might this explain decreased survival among fish and amphibian eggs in these regions?

3. With reference to Question #2 above, why might the effect on bird eggs be **less serious** than that experienced by fish and amphibian eggs?

Self Test

1. On the basis of the results of your experiments, why is it considered dangerous to have an abnormally high fever?

2. The body fluids contain buffers whose function is to minimize changes in pH levels. Why is a buffering system beneficial for optimal enzyme function?

   The following are the results of an experiment that examined the effect of pH on the activity of an enzyme isolated from the human digestive tract.

Tube #	pH	\multicolumn{12}{c}{Elapsed Time (minutes)}
		2	4	6	8	10	12	14	16	18	20	22	24
1	2.0	−	−	−	−	−	−	−	−	−	−	−	−
2	3.0	−	−	−	−	−	−	−	−	−	−	−	−
3	3.3	−	−	−	+	+	+	+	+	+	+	+	+
4	3.5	+	+	+	+	+	+	+	+	+	+	+	+
5	3.7	−	−	−	−	−	−	+	+	+	+	+	+
6	4.0	−	−	−	−	−	−	−	−	−	−	−	−

3. The optimal pH for this enzyme is _____.

4. In a few sentences, summarize the experimental results for pH 3.7.

5. Would you characterize this enzyme as being **general or specific** in regard to its pH requirements? **Explain your answer**.

6. Enzymes regulate the production of the pigment melanin in the fur of Siamese cats and Himalayan rabbits. At normal body temperatures, the rabbits produce white fur, but their paws, ear tips, and noses have black fur. On the basis of your understanding of the effect of environmental conditions on enzyme activity, suggest one possible explanation for the difference in fur color on various parts of the body.

Functions of Tissues and Organs I

Objectives

After completing this exercise, you should be able to:

- list and explain several features present in the epidermis that provide for protection of the body
- explain how each skin layer is different from the others
- identify each structure in the dermis and explain its function
- list and explain at least four examples of how deposits of fats and oils are useful to plants, animals, and microorganisms
- discuss several adaptations of epithelial tissues that demonstrate the relationship between structure and function
- discuss the relationship between the density of touch receptors in various body locations and touch sensitivity
- apply your knowledge of the special features of the skin to practical situations.

CONTENT FOCUS

Within our bodies, we have approximately 100 trillion cells. As you observed in previous exercises, not all cells have the same **structure.** In the bodies of multicellular animals such as humans, not all cells have the same **function.**

A **tissue** is a **group of similar cells that perform a specific function.** Cells are organized into **four major tissue types:**

Epithelial Tissue	Lines all inner and outer body surfaces; covers organs and body cavities
Muscle Tissue	Contracts to produce movements
Connective Tissue	Joins and supports other tissues
Nervous Tissue	Senses stimuli; transmits signals around the body

Tissues group together to form the organs of the body. All of these tissue types are present in the **skin,** the body's largest organ.

To examine tissue function, you will take a closer look at the **design of your arm,** beginning with the outer epidermis and working your way through the underlying tissues to the supporting skeleton.

ACTIVITY 1   SKIN: THE OUTER PROTECTIVE LAYER

1.   Work in **groups**. Set up a **dissecting microscope.**

2.   Place your hand, **palm-side down,** on the stage of the microscope. **Examine your skin** closely, beginning with the **lowest** magnification and gradually zooming up to the **highest.**

      Record your skin **observations** (in detail).

3.   **Remove your hand from the microscope stage.** Get a **dropper bottle of water.**

      While holding your hand level in front of you, **palm-side down (NOT under the microscope),** place a drop of water on the back of your hand.

      What happens to the water droplet?

      Gently **tilt** your hand. **What happens** to the water droplet now?

      Can liquids easily penetrate the skin? _____

      Is this one of the **protective functions** of the skin? _____ **Explain your answer.**

4.   The **waterproofing** quality of the skin is due to the presence of the **protein keratin** in epithelial cells.

      The **keratinized** cells of the epidermis also protect the underlying tissues from **mechanical injuries and abrasions.** On parts of the body where **abrasion is most common,** the epidermal layer tends to be **thicker.**

5. The skin also provides protection from exposure to **ultraviolet (UV) radiation.** Pigment producing cells in the epidermis manufacture the **protein melanin** that blocks penetration of UV rays and protects underlying cells from damage.

    Most people have about the same number of melanin-producing cells, called **melanocytes. Dark-skinned** individuals, however, produce **more and darker melanin** than fair-skinned individuals. Melanin production is also **stimulated by exposure to sunlight.**

6. Problems with melanin production can result in different disorders.

    Albinos have a normal number of melanin-producing cells, but the cells do not synthesize melanin. For this reason, albinos have no pigment in their skin, eyes, and hair. In the condition **vitiligo,** melanocytes die, causing patches of skin to lose their coloration. These light spots in the skin are often surrounded by skin with normal pigmentation.

✔ Comprehension Check

1. Which of the following areas of the body probably have a thick epidermal layer? **(Circle ALL correct answers.)**

    a. inside of the cheeks   d. scalp
    b. palms of the hands   e. elbows
    c. soles of the feet   f. abdomen

2. The thick layers of keratinized cells mentioned above form highly distinctive patterns of **ridges and whorls** in some body locations.

    We call these patterns _____.

3. To raise some money for your tuition, you take a summer job working construction. After several weeks, you notice **calluses** have developed on your hands. What is a **callus? Why did the calluses form?**

4. Based on the information in **Activity 1**, what causes a **suntan?**

5. **Skin cancer** is increasing in our population. **(Circle one answer.)** You have better protection from exposure to the damaging effects of UV rays if you have **dark/fair** skin.

Check your answers with your instructor before you continue.

ACTIVITY 2   THE EPIDERMIS

Preparation

By looking at **Figure 8-1,** you can see that the skin is made up of several layers:

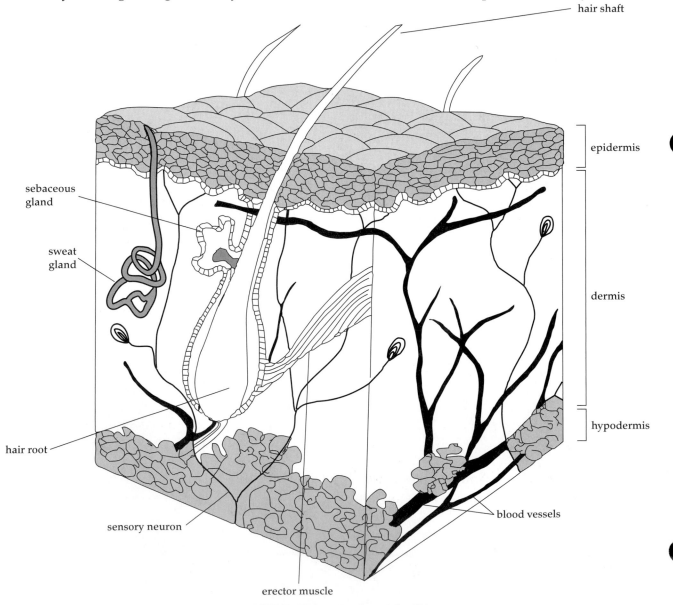

FIGURE 8-1. Layers of the Skin

the **epidermis** (the outermost layer), the **dermis** (the middle layer), and the **hypodermis** (the innermost layer).

1. Look at **Figure 8-2** and note that the **epidermis** consists of many layers of **epithelial cells.**

    You can also see the thickness of the epidermis in comparison to the other skin layers by looking at the top of the diagram in **Figure 8-1**.

    The protective layer of keratinized cells at the **surface** of the epidermis is **repaired by rapidly dividing cells** at the **base** of the epidermis. Newly formed cells are gradually pushed upward to replace lost and damaged cells in the outer layer. In some individuals, excessive shedding of cells occurs in the scalp area.

    These discarded scalp cells are referred to as _____.

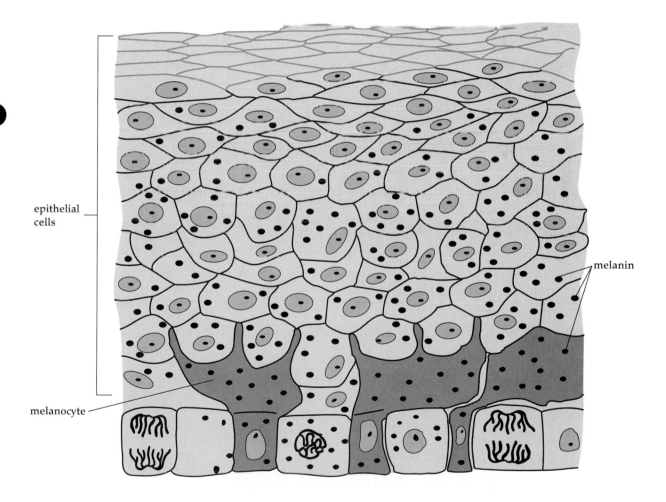

FIGURE 8-2. The Epidermis

✓ Comprehension Check

1. On **Figure 8-2**, draw an **arrow** showing the **direction of cell replacement** in the epidermis.

2. Dividing cells have **distinctly visible chromosomes** within them (these appear as dark lines).

   Using a **colored pencil or highlighter**, color in **two** cells that are dividing.

3. **(Circle one answer.)** Actively dividing cells can be found in the **top/middle/ bottom** layer of the epidermis.

4. On **Figure 8-2**:

   Place an **"O"** inside a cell that would be one of the oldest cells in the epidermis.

   Place a **"Y"** inside a cell that is one of the youngest cells in the epidermis.

   Place a **"B"** inside a cell close to a blood vessel.

5. If the prefix **"epi"** means *above*, where do you predict the **dermis** of the skin will be located?

Check your answers with your instructor before you continue.

ACTIVITY 3   THE DERMIS

1. By looking at **Figure 8-1**, **circle the correct answer** in each of the following statements.

   The dermis is located **above/below** the epidermis.

   The dermis is **thicker/thinner** than the epidermis.

   Hair roots are located in the **dermal/epidermal** layer of the skin.

2. Take a look at the epidermal and dermal layers in **Figure 8-1**.

   Are any **blood vessels** present in the **epidermis?** _____

   If you cut **just** your epidermis, will the cut bleed? _____

3. Blood vessels carry nutrients and oxygen to body cells.

   (Circle one answer.) Epidermal cells at the **base** of the epidermis have **excellent/fair/poor** access to oxygen and nutrients.

   Epidermal cells at the **surface** of the epidermis have **excellent/fair/poor** access to oxygen and nutrients.

   The farther epithelial cells are removed from good access to nutrients and oxygen, the **better/worse** their chances of survival.

4. **Blood vessels** near the root area supply the hair with **oxygen and nutrients** needed for growth.

   Melanin granules in the center of the hair shaft give the hair its color.

   Notice that **sebaceous glands** are located along the hair shaft. Oily secretions from the sebaceous glands **lubricate** the hair and skin, help keep the skin from **drying out**, and **inhibit growth of bacteria.**

Comprehension Check

1. The sale of **hand lotion** is a multimillion dollar market in the United States. Using your knowledge of how the epidermis is lubricated, explain why hand lotion is such a big-selling item.

2. Explain why the skin of a person's scalp is often more oily than the skin of his or her arm.

Check your answers with your instructor before you continue.

ACTIVITY 4   HAIR TODAY, GONE TOMORROW

Preparation

Millions of hairs are scattered over the body. The average scalp has about 100,000 hairs and a man's beard has another 30,000.

1. Refer back to **Figure 8-1**. You will see that a hair consists of two parts, the **shaft** and the **root.** The shaft portion of the hair begins in the **dermis** and extends **outside** the surface of the skin.

    The **root** is found **only within** the dermis and is surrounded by several layers of cells. Collectively, this structure is called a **hair follicle.**

2. Sit with your **eyes closed** while your partner runs his or her hand over the top of your head, **gently** touching the **hairs.**

    Can you tell when you are being touched? _____

    What **section** of the hair was just touched? _____

    A knot of **sensory nerve endings** is wrapped around the **root of each hair.**

    Moving a hair activates these nerve endings, so that your hairs function as sensitive touch receptors.

3. The outer part of the shaft contains heavily **keratinized** cells, which appear roughened and scalelike under a microscope.

    Hair pigment is made by **melanocytes** at the base of the hair follicle. Several types of pigments combine to produce the various hair colors.

    (Circle one answer.) Brown/red/black/blond hair has the most melanin present.

4. Living cells **divide** at the base of the follicle, causing your hair to grow. **Blood vessels** provide follicle cells with the oxygen and nutrition needed for growth.

5. Refer back to **Figure 8-1.** You can see that each hair is connected to a **tiny erector muscle** that **automatically** raises the hair when you are cold.

    In animals with fur coats, raised body hairs **hold a layer of warm air** next to the skin, insulating them from the cold. Humans have too few body hairs for this response to help keep us warm, but we are like other animals in our response to changing temperatures.

    Contraction of the muscles that raise body hairs causes _____ to appear on the surface of the skin.

✓ Comprehension Check

1. Which of the following skin features can be found in the **dermis? (Circle ALL correct answers.)**

   a. melanin   f. hair follicle
   b. blood vessels   g. hair shaft
   c. nerve endings   h. highly keratinized cells
   d. sebaceous glands   i. sweat glands
   e. erector muscles

2. If you get a paper cut that penetrates to the **dermis,** will the cut bleed? _____ **Explain your answer**.

3. If the paper cut penetrates **only** the **epidermis,** will the cut **hurt?** _____ **Explain your answer**.

4. Burns and other injuries sometimes causes the epidermis to separate from the dermis.

   When the space fills up with fluid, we call this a _____.

Check your answers with your instructor before you continue.

ACTIVITY 5   BELOW THE SKIN (THE HYPODERMIS)

1. Beneath the skin is a layer of **connective tissue** with many fat cells. (See **Figure 8-1**.) This layer is called the **hypodermis** ("hypo" meaning **under** the dermis) or **subcutaneous** layer ("sub" meaning *under* and "cutaneous" refers to the skin).

   Perform the following experiment using a **"blubber mitten"** to discover one of the functions of the subcutaneous fat layer.

2. Work **individually.** Get the following supplies: **one blubber mitten and one empty mitten.**

    The two mittens are designed to simulate how the subcutaneous layer of the skin would function **with and without fat cells.**

    In this experiment, you will place one mitten on each hand and immerse both hands in an **ice-water bath.**

3. **Before proceeding,** form a hypotheses about the **effect of water temperature** on each of your hands. **Record** your hypotheses below.

HYPOTHESES FOR BLUBBER MITTEN EXPERIMENT

4. Go to the ice-water bath and **perform your experiment. Record your results** below.

RESULTS OF BLUBBER MITTEN EXPERIMENT	
WITHOUT FAT	WITH FAT

5. From the results of your experiment, what **conclusions** can you draw about the effect of fatty tissue in the subcutaneous layer of the skin?

☑ Comprehension Check

1. Which of the two mittens was the **control** in your experiment? _____

2. Why was it **necessary** to use a **control mitten** (instead of putting your **bare hand** in the ice-bath)?

3. **Based on the results** of your experiment, give **one function** for subcutaneous fat tissue.

4. Again, **referring to the results** of your experiment, can you suggest **two reasons** why some animals eat a lot in the fall season?
   a.

   b.

5. You are probably familiar with the fact that, when you **mix oil and water,** the oil will float to the surface. Do you think that significant **fat** deposits, such as those found in marine mammals, contribute to their buoyancy? **Explain your answer**.

6. On an ocean field trip with your marine biology class, you make a collection of **phytoplankton (tiny marine plants).** When you look at these plants through the microscope, you notice that many of them have **oil droplets** inside their cells.

   Suggest **two ways** in which oil droplets might be helpful to the phytoplankton.
   a.

   b.

7. Often **injections** of medicine are administered into the **subcutaneous layer** of the skin.

   The familiar term _____ **needle** refers to the location in the skin where the needle is inserted.

Check your answers with your instructor before you continue.

ACTIVITY 6   THE SENSORY FUNCTION OF THE SKIN

Preparation

The skin acts as an interface between the body and the outside environment. Information about several aspects of the outside world is acquired through activation of sensory receptors in the dermis and epidermis of the skin.

Sensory neurons process and transmit incoming information to the central nervous system (the brain and spinal cord). Neurons vary in function and appearance, but all share a similar three-part structure: the cell body, the axon, and the dendrites (see **Figure 8-3).**

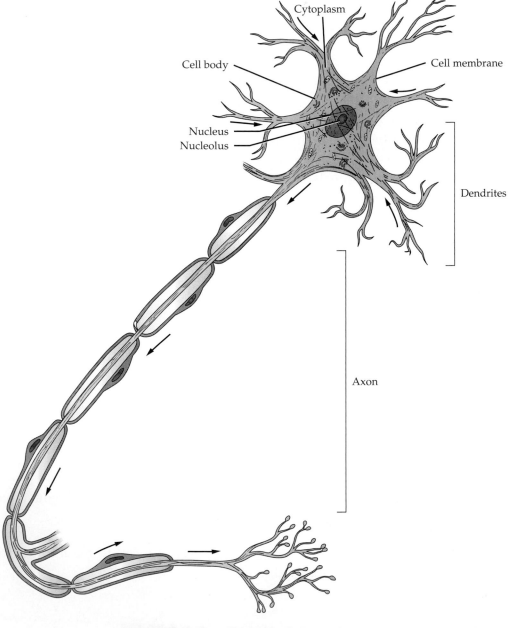

FIGURE 8-3. Structure of a Typical Neuron

Every neuron has a **cell body,** the region of the cell that contains the nucleus and a variety of other organelles necessary for cell metabolism. Outside stimuli are received by **dendrites** in the skin and other sense organs. Incoming signals from these dendrites are transmitted toward the cell body and continue into the **axon.** Axons relay outgoing messages from one neuron to other neurons or to various tissues and organs in the body.

Dendrites in the skin are specialized to receive different types of sensory input, such as **touch, temperature, pressure, pain, vibration, and proprioception** (the sense that tells you the current position of your body and limbs). In the following experiment, you will be making observations about the distribution of touch receptors in the skin.

1. Work **with a partner.** Get the following supplies: **one set of touch calipers.**

    The set of touch calipers consists of a series of corks. Each cork has two pins inserted so that the blunt ends are protruding. The pins are spaced at varying distances from each other (**1mm, 2mm, 4mm, 8mm, 16mm, and 32 mm apart**).

2. Form a hypotheses about the relative sensitivity of various body locations. For the six body locations mentioned in **Table 8-1,** list them in order from the area that you think will be **least sensitive** to the area you think will be **most sensitive.**

TOUCH SENSITIVITY HYPOTHESES

3. Perform the following set of tests on your partner, then reverse roles and have your partner use the same tests to check your touch response sensitivity.

    Your partner should be seated, with his/her **eyes closed.**

    For each touch given, the subject will report whether they feel one or two pins touching the skin.

4. Begin with the **32 mm caliper.** Lightly touch the skin on the back of the test subject's neck.

    Repeat the process with the remaining calipers in the set (**16 mm, 8 mm, 4 mm, 2 mm, and 1 mm**) in any order you wish.

    For each test, record whether the person is able to feel two pins touching the skin.

If the test subject felt **two pins**, record **yes.** If the test subject felt **one pin only**, record **no.**

Record these results in **Table 8-1.**

5. Repeat the same procedures at each of the following locations:

    back of the hand   tip of the nose
    palm of the hand   cheek
    inside of the forearm   tip of the forefinger

6. **Reverse roles** and have your partner repeat the same tests to check your touch response sensitivity. Record the results in **Table 8-1.**

TABLE 8-1
TOUCH SENSITIVITY IN VARIOUS BODY LOCATIONS

TEST SUBJECT 1

BODY LOCATION	32 MM	16 MM	8 MM	4 MM	2 MM	1 MM
Back of neck						
Back of hand						
Palm of hand						
Inside of forearm						
Tip of nose						
Cheek						
Tip of forefinger						

TEST SUBJECT 2

Back of neck						
Back of hand						
Palm of hand						
Inside of forearm						
Tip of nose						
Cheek						
Tip of forefinger						

The **greater the number of touch receptors** in a body location, the **more sensitive** that location will be to touch stimuli. The reason for the increased sensitivity lies in the distribution of the receptors. If the pin points touch **two receptors that are adjacent** to each other, the pins will **activate both receptors together** and it will feel as though you have been touched by **only one pin.**

If the pins touch two receptors that are **not adjacent,** however, each will **fire separately** and you will feel the touch of **both pins.** The ability to feel the two pins is called **"two-point discrimination."**

In locations with a high density of touch receptors, therefore, the skin is much more sensitive to touch and much better at providing information about the number of touches (one pin or two pins). This type of touch sensitivity allows a surgeon or mechanic to perform delicate procedures working by touch alone.

The smallest pin distance when the test subject had effective two point discrimination indicates the body location with the highest density of touch receptors.

The higher the number of touch receptors in a specific location, the greater the probability the pins can stimulate two non-adjacent touch receptors.

Comprehension Check

1. Based on the results in **Table 8-1,** rank all areas tested in terms of their relative sensitivities to touch. List the body locations from the area of **least sensitivity** to the area of **highest sensitivity.**

   Subject 1:

   Subject 2:

2. Which of the tested body locations has the **greatest density** of touch receptors? **Explain your answer.**

3. Were there differences in touch sensitivity between you and your partner? If so, what differences did you observe?

4. Would you expect a decrease in touch sensitivity for each of the following conditions? **Explain each answer**.

   a.   calluses on the palms of your hands:

   b.   scar tissue:

   c.   thick subcutaneous fat layer:

5. People who are visually impaired are able to read books and other printed materials by using the Braille system of writing. In the Braille system, words are represented by patterns of raised dots. Braille readers touch the dot patterns with their fingertips to read the text.

   How does the density of touch receptors in the fingertips make the Braille system possible?

Check your answers with your instructor before you continue.

Self Test

Fill in the blank with the choice that is **most appropriate** to describe the function of each skin structure. Answers can be used **only once.**

a. keratin   f. hair follicle
b. melanin   g. blood vessels
c. erector muscle   h. subcutaneous layer
d. epithelial cell   i. sebaceous gland
e. epidermis   j. dermis

1. ___ Produces oil to lubricate the hair and skin.

2. ___ The layer of skin containing many blood vessels and nerves.

3. ___ Type of protein deposits in the skin that form fingerprints.

4. ___ Type of dead cell that flakes off as dandruff.

5. ___ When the hairs on the back of a dog stand up, this tissue is responsible.

6. ___ Location of fatty tissue that insulates and protects the body.

7. ___ Accumulation of this protein helps protect your skin from the sun's rays.

8. ___ The outermost layer of the skin.

9. We often hear claims made by cosmetic companies that their **lotions** will **"make your skin young again."** Based on the information you have gained in this exercise, do you think that this is an accurate statement? **Explain your answer.**

10. You are determined to quit smoking, so you decide to try a brand of nicotine "patch." Where is the **closest location** for the nicotine to enter your circulatory system? **Explain your answer**.

11. In reference to the nicotine patch: Even though it would not be as visible, why would it be a bad idea to place the patch on the sole of your foot? (Answer in terms of the ability of the patch to **function.**)

12. *Challenge Question!* Burns are classified according to how many skin layers are destroyed. A **first-degree burn** only affects the outer layers of the epidermis. A **second-degree burn** destroys the epidermis and penetrates the dermis. **Third-degree burns** destroy skin tissues down to and including the subcutaneous layer.

    A nurse who works in the burn ward of a local hospital, notices that patients who have second-degree burns suffer more pain than patients who have third-degree burns.

    Using your knowledge of the skin, **explain** why the more severe burn causes less pain.

Functions of Tissues and Organs II

Objectives

After completing this exercise, you should be able to:

- identify three different types of muscle tissue using the microscope
- discuss the functions and body location of each type of muscle tissue
- explain the effects of muscle fatigue on arm muscle activity
- identify the major structures of a long bone and explain the function of each
- discuss the contribution of minerals and protein fibers to bone strength and flexibility
- apply your knowledge of bone and joint structure to practical examples of skeletal support and movement.

CONTENT FOCUS

In previous exercises you looked at the protective outer covering of the body. Now you will move beneath the skin to examine muscles and bones.

Consider an Olympic sprinter at the beginning of a race. The moment the starting gun fires, muscles contract, the skeleton responds, joints allow for changes in body position, and tissues all over the body work together to coordinate forward movement. Tasks as simple as sitting, standing, walking, or taking notes in class all require similar coordination of muscles, bones, and joints.

ACTIVITY 1   BETWEEN THE SKIN AND BONES

1. If you were to take a **small piece of steak** and make a wet mount of the **muscle tissue,** you would see alternating light and dark bands. These bands or **striations** are composed of two proteins **(actin and myosin)** which are involved in **muscle contraction.**

2. To observe bands of actin and myosin, get a **prepared slide of skeletal muscle** (also called **"striated,"** meaning striped, muscle).

3. Observe the slide with the **compound microscope** on **high power.** Refer to the **photograph** in **Figure 9-1** to locate the following structures on your slide:

   a. **one** muscle cell (also called a muscle **fiber**)

   b. striations

   c. nuclei of **one** muscle cell

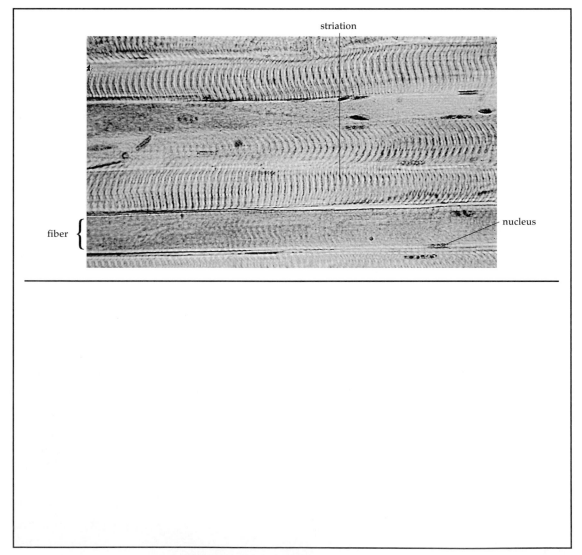

FIGURE 9-1. Skeletal Muscle Tissue

4. In the empty box in **Figure 9-1,** make a drawing of several skeletal muscle cells **as they appear in your microscope.**

5. **Label** the **striations, nuclei, cytoplasm, and cell membrane** of **one** skeletal muscle cell on your **drawing.**

Check your answers with your instructor before you continue.

6. The **function** of skeletal muscle is to bring about **voluntary** movement. Skeletal muscles are under your conscious control. **Wiggle the fingers** of your left hand. You are using skeletal muscles!

7. There are **two other types of muscle tissue** that have **different functions** in the body. **Smooth muscle** allows for **involuntary** movements—that is you have little control over its function.

   Observe the **two drawings** of smooth muscle in **Figure 9-2.**

   The protein fibers of smooth muscle cells are **not** arranged in bands, so **striations are not visible** in this tissue. Smooth muscles are found in the **digestive tract, blood vessels,** various **glands,** and many other locations where **automatic responses** are required.

8. **Add labels** for the following structures to the drawing of **individual smooth muscle cells** in **Figure 9-2: cell membrane, nucleus, and cytoplasm.**

9. **Cardiac muscle** is found **only** in the **heart.** It is responsible for your heartbeat. Refer to the photograph and drawing in **Figure 9-3.**

   Are cardiac muscle cells **striated** (similar to skeletal muscle cells)? _____

10. Contraction of all cardiac muscle fibers **must be coordinated** in order to maintain a regular heartbeat.

   To accomplish this coordination, cardiac muscle cells are tightly connected into a communication network that sends the signal to contract from cell to cell. These cell junctions are called **intercalated discs.**

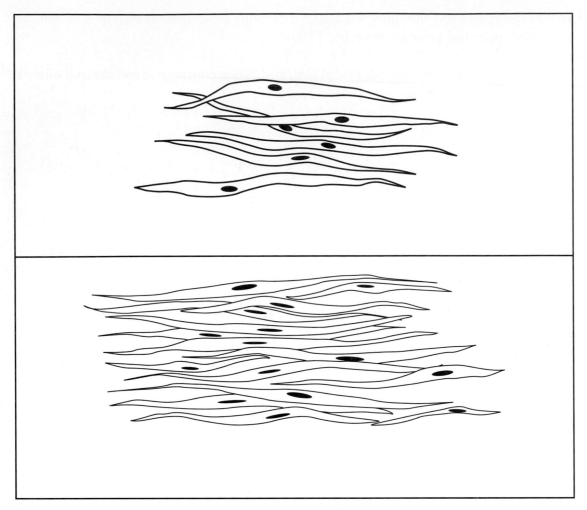

FIGURE 9-2.   Smooth Muscle Tissue

✓ Comprehension Check

Now that you are familiar with the **three different types of muscle tissue,** answer the following:

1. The erector muscles that **raise hairs on your arm** consist of _____ muscle tissue.

2. When you **wink** your eye, what type(s) of muscle tissue(s) is(are) in use?

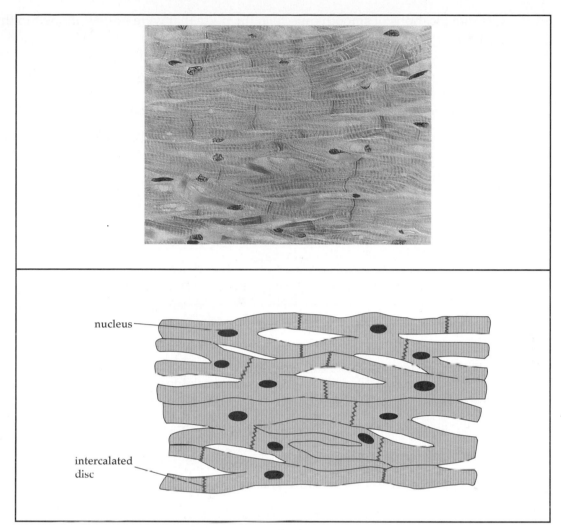

FIGURE 9-3. Cardiac Muscle Tissue

3. When you are overheated, extra blood is directed to the skin for cooling. What type(s) of muscle tissue is(are) involved?

4. During **labor and childbirth,** what type(s) of muscle tissue(s) is(are) contracting? **Explain your answer**.

5. List **two** locations of smooth muscle in the body that were **not** mentioned in this exercise.

    _____   _____

6. During a heart attack, _____ **muscle** may be damaged.

 Check your answers with your instructor before you continue.

ACTIVITY 2   MUSCLE FATIGUE

Preparation

Muscles that are actively contracting become weaker as time passes. This process is known as **muscle fatigue.** Muscle fatigue can be explained by several factors which occur simultaneously as your body works. These include:

■ lack of ATP to meet energy needs

■ insufficient oxygen for cell respiration

■ depletion of energy reserves in the muscle cells

■ accumulation of metabolic wastes (such as lactic acid)

1. Work in groups. During this experiment, you will be supporting a **heavy** book on your open palm with your arm completely extended. Which arm can support the book longer? **Form a hypotheses** about the ability of the muscles in your right and left arm to support the book and **record** your hypotheses below.

HYPOTHESES FOR MUSCLE FATIGUE EXPERIMENT

Check your hypotheses with your instructor before you continue.

2. For your muscle fatigue experiment, one member of the group will act as a **time-keeper**, the second will **record** the experimental results, and the third will be the **test subject**.

Note:

The person holding the book should not know how much time has expired until the entire experiment is completed.

---

3. Get an **extremely heavy book**.

   a. Place the book on your open hand.

   b. With the book on your hand, extend your arm fully with the palm up. **Do not bend your elbow.**

   Your arm should **not** be braced against your body.

   c. Record the length of time you can hold your arm extended. **Record the results** in **seconds** under **Trial 1** in **Table 9-1**.

   d. Rest your arm for **5 seconds** and repeat the experiment for **Trial 2**.

   e. Rest your arm for **5 seconds** and repeat the experiment for **Trial 3**.

   f. Rest your arm for **5 seconds** and repeat the experiment for **Trial 4**.

TABLE 9-1		
RESULTS OF MUSCLE FATIGUE EXPERIMENT		
Trial Number	Time Arm Held Extended (SEC) Right Arm	Time Arm Held Extended (SEC) Left Arm
1		
2		
3		
4		

4.   **Graph** your experimental results in **Figure 9-4.** Plot **time** on the **Y axis.**

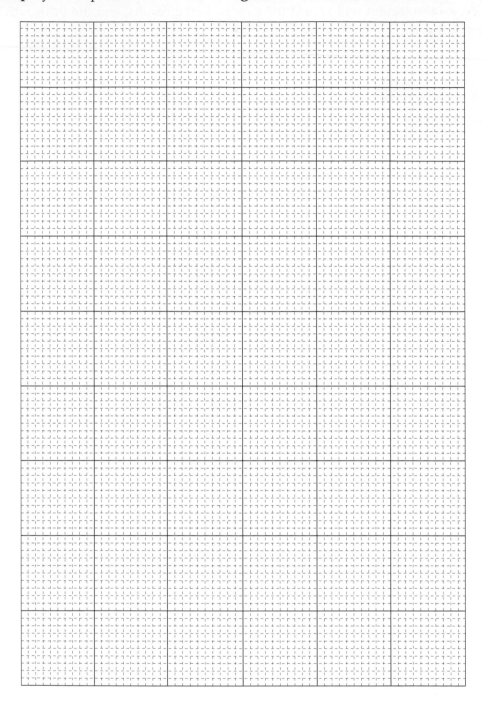

FIGURE 9-4.   Comparison of Muscle Fatigue in Right and Left Arm

5. Did the experimental results **support** your hypotheses? _____

    Explain your answer, mentioning **facts** collected during your experiment.

6. Did you see evidence of **muscle fatigue? Explain** your answer.

7. To have full confidence that your conclusions about muscle fatigue are accurate, **what changes would you recommend** for the experimental design?

 Check your answers with your instructor before you continue.

ACTIVITY 3   SKELETAL SUPPORT

Preparation

Bone is a type of **connective tissue.** Although bone may not superficially resemble other body tissues, it **is** composed of **living cells.**

Bone cells produce a rigid material that forms the structure generally associated with bones. This mineral **"matrix"** in humans is composed of **calcium phosphate.**

Interwoven throughout the rigid mineral matrix are **protein fibers** that add **strength and flexibility.**

Compact bone is strong and densely packed. In compact bone, the cells are organized into vertical structures that resemble **drinking straws.** If you try to crush a straw by pushing on both ends, it will be hard to do. In a similar manner, bones resist the force of gravity and support your body weight.

In contrast to compact bone, **spongy bone** consists of a loose arrangement of thin plates of bone similar to the appearance of a sponge. **Spongy bone is lighter** than compact bone, **reducing the weight of the skeleton** and making it easier for the muscles to move the bones.

Bone strength is a result of compact and spongy bone working together. **Compact bone** can only resist forces applied **along the length** of the bone, while **spongy bone** helps resist stresses **applied from many directions** (such as a kick to the side of your leg).

The inside of a bone contains **bone marrow.** The central cavity of the bone **shaft** contains **yellow marrow,** which functions as a fat-storage area. At the **ends of the long bones,** gaps in the spongy bone contain **red marrow,** which produces all types of **blood cells.**

Bone cells, as with cells of all tissues, require delivery of oxygen and nutrients and removal of waste materials. Although you cannot see them without magnification, **channels throughout the bone** provide passageways for a network of blood vessels and nerves to serve these cells.

1. Work in groups. Get a **long bone** that has been cut open lengthwise.

   Set up a **dissecting microscope** for your observations.

2. **Examine** the bone with and without the microscope.

   On the basis of your **observations,** label the following structures in **Figure 9-5: compact bone, spongy bone, shaft, marrow cavity, the location of red marrow, and the location of yellow marrow.**

3. **(Circle one answer.)** Most blood cells are produced in the **shaft/ends** of the bone.

   Explain your answer.

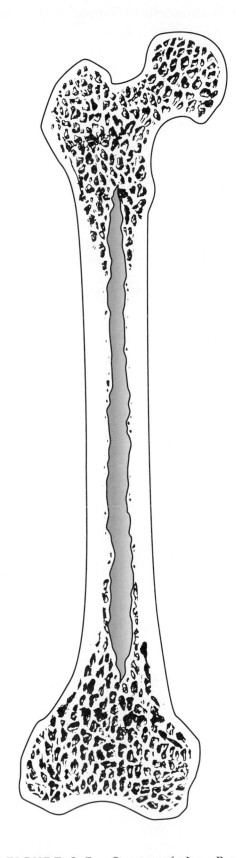

FIGURE 9-5. Structure of a Long Bone

4. **(Circle one answer.)** In situations of long-term starvation, **red marrow/yellow marrow** would probably decrease first.

   Explain your answer.

5. Examine a **human pelvis,** which is composed of several attached bones.

   Based on the function of the pelvis and its position in the skeleton, form a hypotheses about the relative **amounts of compact and spongy bone** present in the bones of the pelvis. **Record** your hypotheses below.

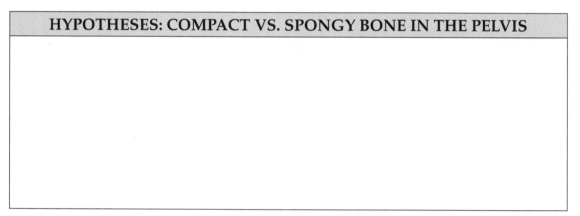

HYPOTHESES: COMPACT VS. SPONGY BONE IN THE PELVIS

6. **Explain** the line of reasoning you followed to develop your hypotheses.

7. Suggest a method you could use to **test** your hypotheses.

Check your answers with your instructor before you continue.

8. *Challenge Question!* In many societies, heavy bundles are balanced on people's heads, rather than carried in their arms or on their backs (see **Figure 9-6**).

   Explain how **compact and spongy bone** each contribute to the ability of the skeleton to support the load.

FIGURE 9-6. Supporting a Load on Your Head

ACTIVITY 4
WEAK AND STRONG BONES: CLASSROOM DEMONSTRATION

Bone is about 75% **minerals (calcium phosphate)** and 25% **protein fibers (mostly collagen).** These materials, **acting together,** give bone its strength and flexibility.

1. Your instructor has prepared several **chicken bones** that have been subjected to **three different treatments.** The bones will be used to demonstrate how **minerals and protein fibers** in bone tissue contribute to strength and flexibility.

   Dish A Contains cleaned chicken bones that were soaked in acid for several days. Under these conditions, **minerals** are removed, but the **protein** content does not change.

   Dish B Contains cleaned chicken bones that were baked in a very hot oven. Under these conditions, the **protein fibers** are destroyed, but the **mineral** content does not change.

   Dish C Contains chicken bones that were cleaned and dried.

2. Which dish contains the **control** bones? _____

3. For each statement, **circle one answer**.

   The acid-treated bones are **more/less flexible** than the control bones.

   The acid-treated bones are **harder/softer** than the control bones.

   Explain your answer to both statements.

4. For each statement, **circle one answer**.

   The baked bones are **easier/harder** to **break in half** than the control bones.

   The baked bones are **easier/harder** to crush **into powder** than the control bones.

   Explain your answer to both statements.

✓ Comprehension Check

1. In the condition known as **rickets,** the leg bones bend under the body's weight. Using your knowledge of the role of **minerals and protein fibers** in bone structure, explain one possible cause of rickets.

2. In the condition **osteoporosis,** the spongy bone of the pelvis and femur (long bone of the thigh) can be affected by severe loss of bone mass. Using your knowledge of bone structure, explain why a person with this disease would be more susceptible to hip fractures.

Check your answers with your instructor before you continue.

ACTIVITY 5
A JOINT VENTURE: MOVING THE BONES OF THE SKELETON

Preparation

Since bones are rigid, the skeleton can only change position through the action of joints. Most joints of the body are **synovial joints.** At synovial joints, the adjacent bones are separated by a fluid-filled capsule. Some joints, such as the knee, give us more problems than others. The problems originate because of weakness in the joint supports.

In the hip and shoulder joints, a **ball** on the end of one bone fits securely into a **socket** on another bone. The knee joint, in contrast, is held together only by tough bands called **ligaments.** The ligaments stretch like thick rubber bands between the ends of the thigh and shin bones, pulling them toward one another.

The **collateral ligaments** support the joint on the right and the left sides. The **cruciate ligaments** cross inside the knee from front to back, preventing the knee from overextending.

1. Work in groups. Get a **knee model.**

2. Locate the **two collateral and two cruciate ligaments.** Ligaments form attachments **between bones. Label** the **four support ligaments** on **Figure 9-7.**

3. The knee joint connects the **femur** (thigh bone) and the main support bone of the lower leg, the **tibia** (shin bone). The smaller lower leg bone, the **fibula,** is not part of the knee joint. Although the fibula does not help support body weight, it is an important surface for muscle attachment and provides stability to the ankle joint.

   Label the three leg bones on **Figure 9-7.**

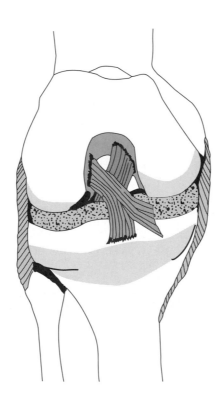

FIGURE 9-7. Synovial Joint of the Knee

4. The **patella** (kneecap) protects the joint and prevents the knee from bending in the wrong direction.

5. The **meniscus,** a thick pad between the femur and tibia, **improves the fit** between the bone ends, making the joint more stable. Locate the **meniscus** on the knee model and label it on **Figure 9-7.**

6. Muscles and bones are connected by **tendons.** The ligaments and tendons of the knee form a **capsule** around the joint (see **Figure 9-8).**

7. No matter how smooth the ends of the bones may look, their surfaces are still uneven. When placed close together, these uneven surfaces rub against one another, producing heat, friction, and wear. In joints, the ends of the bones are covered with a **cushioning layer** of **articular cartilage.**

8. A membrane that lines the joint capsule, called the **synovial membrane,** produces **fluid** that lubricates the joint (called **synovial fluid**), reducing friction.

9. Although they are not present on the model, the knee joint also has many fluid-filled sacs called bursae (singular **bursa**).

   These tiny bags of lubricating fluid reduce rubbing in areas where bones come in contact with ligaments, muscles, and skin.

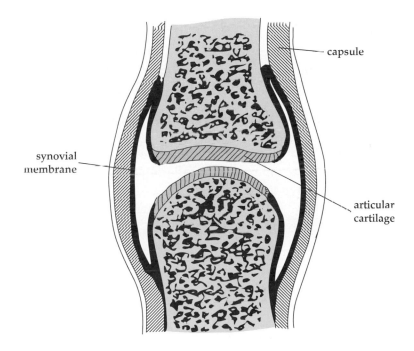

FIGURE 9-8. Interior of a Synovial Joint

✓ Comprehension Check

Fill in the blank with the **most appropriate** answer. Answers can be used **only once**.

a. meniscus
b. collateral ligament
c. articular cartilage
d. synovial membrane

e. synovial fluid
f. cruciate ligament
g. bursae
h. patella

1. ___ Covers the rough ends of bones, creating a smooth surface.

2. ___ A frequent football injury occurs when a player is hit on the side of the knee, tearing this structure.

3. ___ When this structure tears, there is nothing to prevent the knee from extending too far forward.

4. ___ If this structure, which provides a cup for the ends of the long bones, is damaged, the bones may slip out of position in the joint.

5. ___ Excess "fluid on the knee" is produced by this structure.

6. ___ Repetitive movement may lead to inflammation of these bags of cushioning fluid, producing swelling and pain.

Check your answers with your instructor before you continue.

Self Test

Fill in the blank with the choice that is **most appropriate** to describe the function of each in the body. Answers can be used **MORE THAN once.**

a. spongy bone
b. skeletal muscle
c. cardiac muscle
d. yellow marrow

e. red marrow
f. smooth muscle
g. compact bone
h. shaft

1. ____ A storage tissue for fat in bones.

2. ____ Tissue that produces blood cells.

3. ____ Densely packed tissue found at the outside of an arm bone.

4. ____ When the hairs on the back of a dog stand up, this tissue is responsible.

5. ____ When an emergency room doctor use an electrical shock to start someone's heart, they are stimulating this type of muscle.

6. ____ When you touch a hot stove and jerk your hand away, this muscle type is contracting in your arm.

7. ____ This type of muscle moves the leg at the knee joint.

8. ____ Skeletons of birds adapted for flight would have more of this type of bone.

9. ____ This part of a long bone will be longer in taller people.

10. ____ If you placed an arm bone in a vise and applied pressure to both ends, this type of bone would resist breaking.

11. You and a friend are at the gym working out. Your friend is upset because she can't do as many repetitions of a bench press in her third set as she did in her first set. What would you tell her?

12. You are a medical examiner collecting bones from a building where there was a very intense fire. You find that many of the bones are brittle and crush easily. **Explain** this observation.

13. **(Circle one answer.)** The upper end of the patella is connected to a large muscle in the thigh by a **tendon/ligament.**

     The lower end of the patella is connected to the tibia by a **tendon/ligament.**

Introduction to Anatomy: Dissecting the Fetal Pig

Objectives

After completing this exercise, you should be able to:

- identify and compare the external anatomical features of the male and female fetal pig
- identify and explain the function of each of the major structures of the mouth, the neck, and the thoracic cavity
- explain how organ structures are specialized to perform specific functions and give examples
- compare and contrast pig anatomical features with human anatomical features
- discuss the difference between lung capacity with normal and deep breathing.

CONTENT FOCUS

During the next several weeks, you will be studying mammalian anatomy and physiology. To help you get a clear idea of the various organ systems and how they work, you will be looking at the anatomy of the **fetal pig.**

Why fetal pigs? It may surprise you to learn that humans and pigs are very similar in anatomy. Today, pig organs and skin are frequently used for transplants and grafts to replace damaged human tissues. In addition, the fetal pig skeleton is not fully calcified, making dissection easier to perform. During extreme weather, farmers often sell surplus animals rather than run the risk of them dying from heat or cold. As part of the butchering process, all the organs, including the uterus, are removed. If fetal pigs are found in the uterus, they are preserved for educational purposes.

ACTIVITY 1   GETTING STARTED

1. Get the following supplies: **two long pieces of string, a tray, several paper towels, and some dissecting instruments (one scalpel, one pair of scissors, one large pair of forceps, and one blunt probe).**

2. Your laboratory instructor will distribute fetal pigs to each group or give you instructions for obtaining a pig.

3. Spread **two paper towels** on the tray. **Place the pig on its back on the tray.**

4. Tie the **ends** of the first string **tightly** around the two front hooves of your pig and slide it **underneath the tray.** Tie the **ends** of the other string around the two rear hooves and slide it underneath the tray.

5. Rotate the tray so that the pig's **tail is facing you.**

    Where is the **left side of the pig?** _____

> ### Note:
>
> Keep the position of the pig in mind as you proceed with the dissection. The instructions refer to the **pig's right and left sides, NOT your right and left sides**.

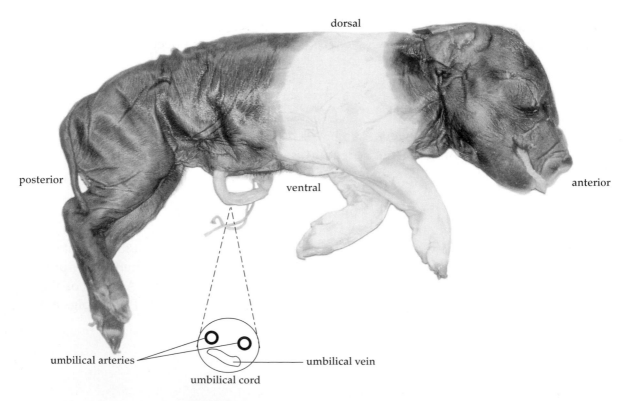

FIGURE 10-1.   Anatomical Locations in the Pig

ACTIVITY 2   FOLLOWING ANATOMICAL DIRECTIONS

Preparation

Before we begin the dissection, it is necessary to understand some special terms that relate to anatomical locations and directions in the pig's body. Using **Figure 10-1** as a guide, use the correct anatomical terms to describe the location of the following:

1. The pig, as placed on the tray, is lying on its _____ surface.

2. The umbilical cord is located on the _____ side of the pig.

3. The attachment of the umbilical cord to the belly is _____ to the **hind** legs.

4. The attachment of the umbilical cord to the belly is _____ to the **front** legs.

ACTIVITY 3   EXTERNAL ANATOMY OF THE FETAL PIG

1. Determine the **sex** of your pig, using **Figures 10-2 and 10-3** as a guide. There are two ways to tell males from females **externally.**

> ### Note:
> **Make sure you observe pigs of both sexes.**

2. Look for the **anus,** which is located close to the base of the tail.

   Female pigs have a small, fingerlike projection, **just ventral to the anus.** This projection is called the **genital papilla.** On either side of the papilla, are the two **labia.** The **urogenital opening** is located between the two labia. The term **urogenital** refers to the double function of this opening. It allows for the excretion of urine ("uro" refers to the urinary system) and also leads to the **reproductive structures** ("genital" refers to the reproductive system).

> ### Hint:
> **If the papilla is not evident, you have a male pig.**

3. In **male pigs,** you will notice the skin appears puffy or baggy just **anterior to the anus.** This is the **scrotum,** which contains the **testes.**

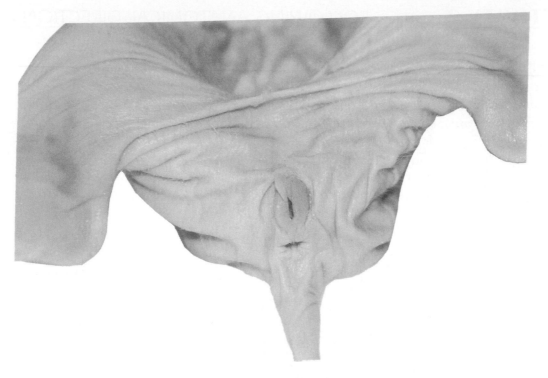

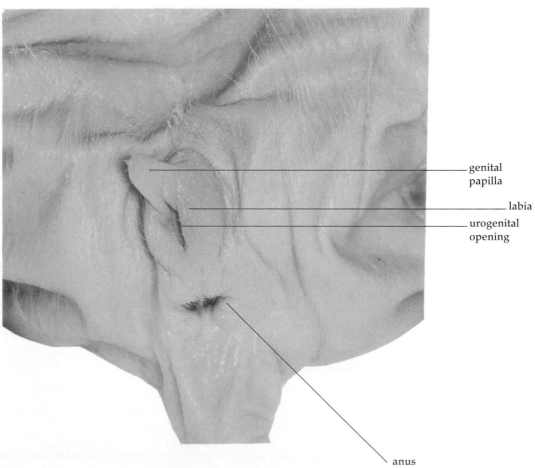

genital
papilla

labia

urogenital
opening

anus

FIGURE 10-2. Female External Anatomy

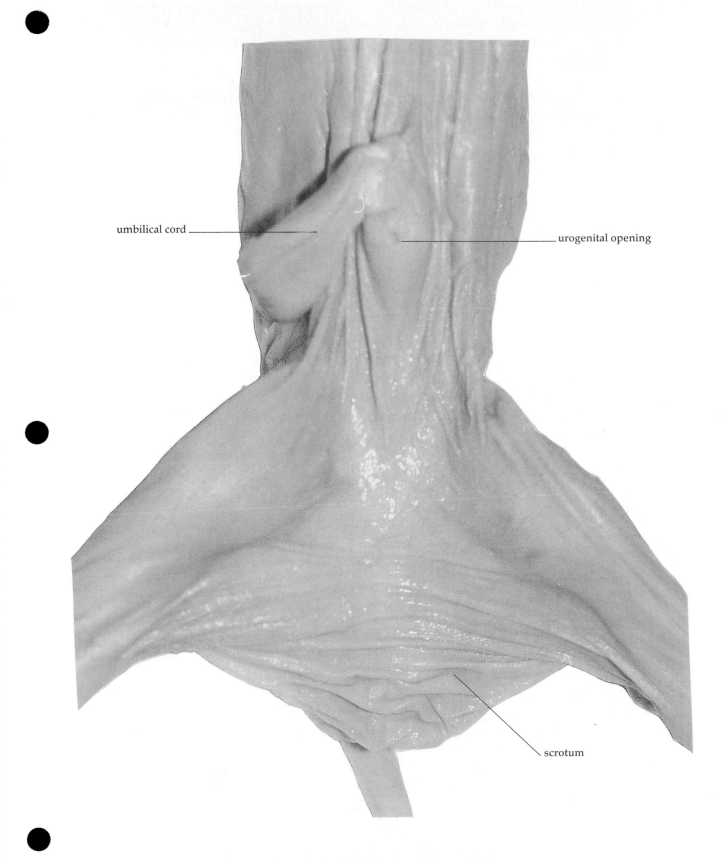

FIGURE 10-3. Male External Anatomy

4. The **urogenital opening** in the male is located just **posterior** to the attachment of the **umbilical cord** to the body wall. As with the female pigs, this opening functions in both the urinary and reproductive systems. Both urine and sperm are released here, but not at the same time.

    You cannot see the **penis** on the exterior of the body because pigs, like many other four-legged animals, have a **retractable** penis, which can be seen when the animals reproduce. You will notice that both male and female pigs have two rows of **nipples** on either side of the umbilical cord. As with humans, the nipples are only functional in females.

5. The body of the pig is divided into three regions: **head, neck,** and **trunk.** The trunk region, with the four legs and tail attached, is further subdivided into the **anterior** area, the **thorax** (chest area), and the **posterior** area, the **abdomen.** The thorax contains the heart and the lungs, enclosed by a protective rib cage. The abdomen contains the organs of the digestive system along with organs of many other important systems.

    The **umbilical cord** attaches the fetus to the **placenta** in the mother.

6. Each pig has a **slit** cut in the side of the neck. This cut was made to inject the circulatory system with **colored latex** (a form of liquid rubber). The **arteries** have been injected with **pink latex** and the **veins** have been injected with **blue latex.** This will help you tell the blood vessels apart when you study the circulatory system.

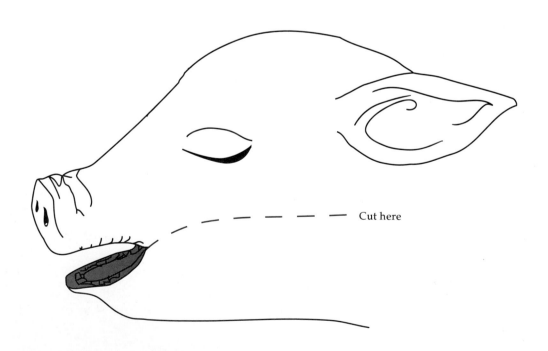

FIGURE 10-4. Pattern for Jaw Incisions

ACTIVITY 4   DISSECTION OF THE MOUTH

1. Gradually, cut through **both sides of the jaw,** as illustrated in **Figure 10-4.** As you cut, **alternate** between the right and left sides until you can open the mouth wide.

Caution!
Young pigs have very sharp, pointed teeth. Be careful where you put your fingers while opening the mouth.

2. Look at the **roof** of the mouth and **feel its texture.** The rippled **anterior region** is hard because it is composed of **bone and cartilage.**

   This area is called the **hard palate.** Notice that the hard palate is divided into **two halves.** The line separating the left and right sections of the hard palate runs directly along the **body midline.**

   Using a pencil, **label** the **hard palate** and the **midline** on the diagram in **Figure 10-5.**

3. The **posterior region** of the roof of the mouth is the **soft palate.** Feel the texture of this area. **How is it different** from the hard palate area?

4. The hard and the soft palate separate the mouth from the **nasal cavity.** As you may have experienced, drainage from the nose can enter the mouth, so there must be a **common area** where these two regions come together. This is the **pharynx.**

   The pharynx can be seen **posterior** to the soft palate.

   Insert the **tip of your blunt probe** into the opening of the pharynx and probe **anteriorly underneath the soft palate. Your probe is now in the nasal cavity.**

   Using a pencil, **add labels** for the **pharynx** and **soft palate** to **Figure 10-5.**

5. Pull the mouth completely open and look down into the throat. There is a hood-shaped flap of tissue called the **epiglottis.** The epiglottis prevents food and liquid from entering the **glottis,** the opening to the air passageways (the prefix **"epi"** means *outside*).

   Using a pencil, **add labels** for the **epiglottis** and **glottis** to **Figure 10-5.**

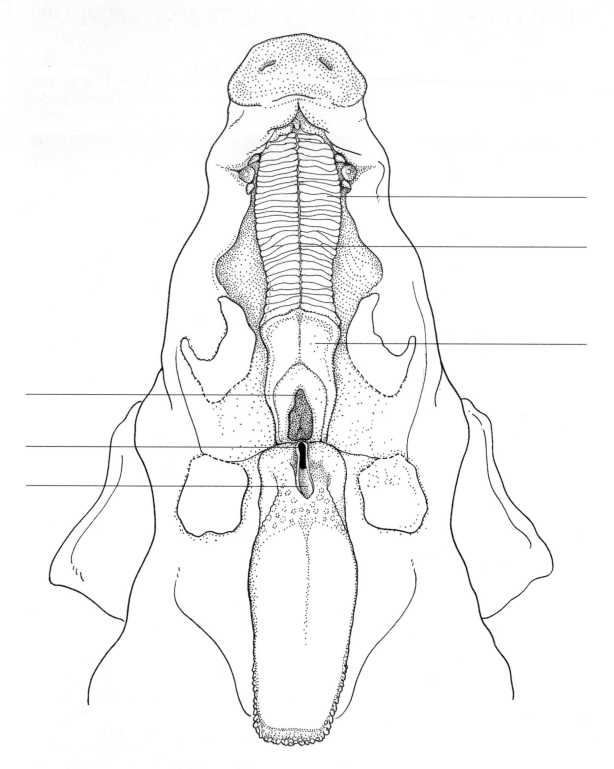

FIGURE 10-5. Structures of the Mouth

Check your Figure 10-5 labels with your instructor before you continue.

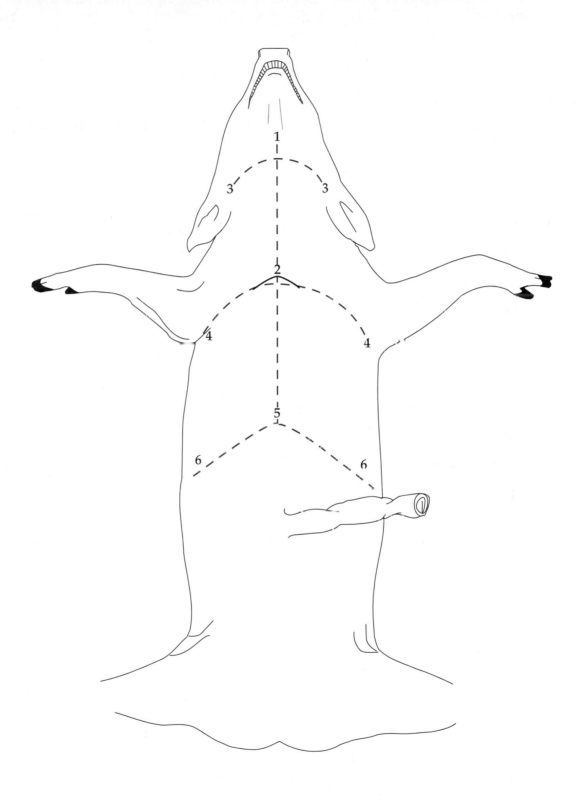

FIGURE 10-6. Pattern for Body Cavity Incisions

6. If you are eating and your food "goes down the wrong pipe," **what pipe** did the food enter? _____

ACTIVITY 5   DISSECTION OF THE NECK

Caution!
While cutting, be careful to keep the point of the scissors away from underlying organs.

1. **Using the dotted lines on Figure 10-6** as a guide, use **scissors** to cut through the **skin and muscle layers.** Begin at the **chin whiskers (number 1)** and continue to the **anterior part of the sternum or breastbone (number 2).**

2. Make two additional incisions **across the throat (number 1 to number 3).**

3. Use your **blunt probe** to separate the muscles from the organs in the neck region. You will expose the **thymus gland** on both sides of the neck. The thymus gland has an important role in **immune system function.**

4. With your fingers, feel around at the **anterior end** of the dissected area **(toward number 1 on Figure 10-6),** until you locate a hard bulblike structure. This is the **larynx.**

    Remove the muscle tissue that is covering the larynx until it can be clearly seen. The larynx is sometimes known as the **"voice box."**

    Considering the **function** of the larynx, what vibrates inside this structure? _____

5. Continue to expose the structures connected to the **posterior end of the larynx.** Remove the overlying tissues until you can see the tubular **trachea,** commonly known as the **windpipe.**

6. The trachea is supported along its length by many **cartilage rings.** The trachea's structure, with the cartilage rings, is similar in design to a vacuum cleaner hose.

What **function** do the metal rings serve in the vacuum cleaner hose?

What is the **function of the trachea** in your body?

Why is the **presence of cartilage rings** in the trachea important for your survival?

7. At the **posterior end of the trachea,** you will see a small, dark-brown organ, the **thyroid gland.** The thyroid gland produces **thyroid hormone,** which controls the body's **metabolic rate.**

8. Using your **blunt probe,** carefully separate the connective tissue from the sides of the trachea. The esophagus is **dorsal to the trachea.**

   Feel underneath the trachea and **lift the esophagus** with your probe. How is the esophagus **different in structure** from the trachea?

9. Is the esophagus part of the respiratory system? _____ **Explain your answer.**

Check your answers with your instructor before you continue.

10. On the side of the neck **opposite to the slit,** carefully remove the overlying muscle and connective tissue to expose the **carotid artery** and **two jugular veins.**

    These vessels will be running **parallel** to the trachea (see **Figure 10-7**).

11. The **carotid artery** is the **closest to the trachea** and should be filled with **pink** latex. The **two jugular** veins should appear **blue**. The carotid artery carries blood rich in oxygen and nutrients toward the head. The jugular veins carry deoxygenated blood back toward the heart.

12. *Dissection Challenge!* Can you find the **vagus nerve** in the neck? It looks like a **flat,** white thread running close to the carotid artery and internal jugular vein. The vagus nerve is crucial for maintaining the normal state of homeostasis. Among other functions, this nerve **regulates heart rate, breathing, and digestive system activity.**

13. **Label** the following structures on **Figure 10-7: thymus, larynx, trachea, thyroid gland, carotid artery, and jugular veins.**

Comprehension Check

Fill in the blanks with the choice that is **most appropriate** for the listed functions. Answers can be used **only once.**

a.	Carotid artery	d.	Trachea	g.	Thymus
b.	Esophagus	e.	Thyroid gland	h.	Jugular vein
c.	Larynx	f.	Vagus nerve		

1. _____ Tube that transports food to the stomach.

2. _____ A blood vessel carrying deoxygenated blood back to the heart.

3. _____ This structure produces a hormone that controls the body's metabolic rate.

4. _____ Tube held open by cartilage rings.

5. _____ A blood vessel carrying in nutrient- rich blood to the head.

6. _____ This structure contains the vocal cords.

7. _____ Plays an important role in immune system defense.

8. _____ Regulates heart rate, breathing, and digestive system activities.

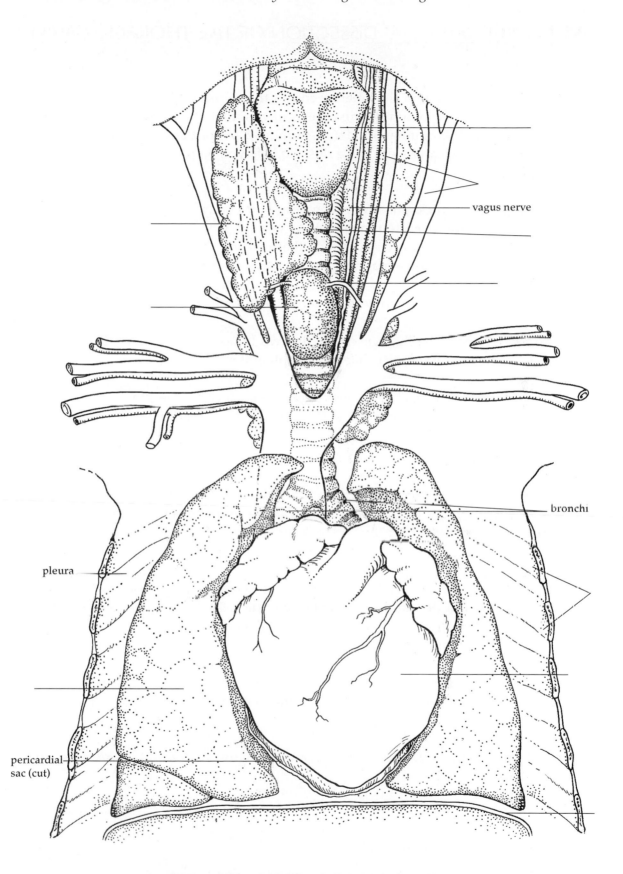

FIGURE 10-7. Organs of the Neck and Thoracic Cavity

ACTIVITY 6   DISSECTION OF THE THORACIC CAVITY

1. With your fingers, find the **anterior end of the sternum.** With your **scissors,** cut through the **sternum and the rib cage** until you reach the end of the rib cage (cut from **number 2 to number 5** and from **number 2 to number 4** on **Figure 10-6**). This will give you a clearer view of the ribs and help you pinpoint the site of your next incision.

2. **Make two more incisions** toward the sides of the body to completely open up the chest cavity **(number 5 to number 6).** As you cut, you will be separating the **ribs** from the **diaphragm,** a large muscle that extends **completely across** the body cavity. **The diaphragm expands the chest cavity so that air flows into the lungs.**

    The thoracic cavity is lined with a smooth membrane.

3. As you look into the thoracic cavity, the **two lungs** are easy to see. Each lung is enclosed in a **pleural cavity,** lined by the **pleural membranes**.

    The lungs are composed of millions of tiny sacs which are the site of gas exchange for the whole body. **Oxygen enters the lungs and carbon dioxide is removed.**

Note:

The trachea extends posteriorly from the larynx and divides into branches called bronchi, which extend to the lungs. The branches cannot be seen at this stage of the dissection, but we will examine them later.

---

4. Located between the two lungs is the **heart.** The heart is a muscular organ that **pumps blood to the lungs and the body tissues.** The pig heart is located in the center of the chest cavity. This is also true of the human heart (not on the left side as is commonly believed).

    The **pericardium** (also called the **pericardial sac**) is a tough membranous sheet that completely encloses the heart. **Remove the pericardial sac** so that you can get a better look at the heart.

5. **Add the following labels** to the diagram in **Figure 10-7: lungs, heart, ribs,** and **diaphragm.**

Check your Figure 10-7 labels with your instructor before you continue.

6.   At the completion of your dissection, do the following:

   a. Get a plastic storage bag for your pig. Place a strip of tape near the bottom of the bag and write your group's names on the tape.

   b. Slide the pig off the tray with the string still attached to the limbs.

   c. Wrap the pig in damp paper towels and seal it in the storage bag. Your instructor will tell you where to store your pig.

   d. Dispose of the remaining pig tissues as instructed. Wash and dry all instruments and trays.

Clean your trays and tools. Your instructor will give you specific instructions for disposing of any remaining pig tissues.

ACTIVITY 7   MEASURING LUNG CAPACITY

Preparation

Just looking at the lungs does not give you a very good idea of how the lungs function. The following activities will allow you to experiment on yourself and your lab partners and develop some ideas about how respiratory function adjusts to your body's needs when you exercise.

The amount of air that enters and leaves your lungs each time you take a breath is called the **tidal volume.** In most men and women, the tidal volume is about **500 cc (half a liter).**

You can increase the amount of air inhaled and exhaled by deep breathing. The **maximum** amount of air that you can move in and out during a single breath is called the **vital capacity.**

1.   Work **with a partner. Each** student will perform this experiment and record their **individual** results.

Caution!
DO NOT use the balloon method if you have respiratory or heart problems!

2.   Get a metric ruler, a clamp, an index card, and a balloon for each person.

Note:

Read through the directions COMPLETELY before beginning!

---

3.   Measure your respiratory volume according to the following directions.

   a. Sit down and relax.

   b. **Inhale normally** and then **exhale** into the balloon.

       Exhale as much air as you can with only one breath.

Note:

If you have trouble inflating the balloon, stretch it several times and try again.

---

   c. **Immediately** twist the balloon several times and **clamp it shut** so that no air escapes.

   d. Place the balloon on its **side** on the laboratory counter. Use the ruler to measure the height of the balloon **in centimeters.**

       Place the ruler near the **center** of the balloon. Don't push down on the balloon when you're measuring.

       An easy way to do this is to place the ruler standing up behind the center of the balloon. Place an **index card** lightly on top of the balloon. Where the **index card meets the ruler,** take your height measurement.

   e. **Record** your results and **repeat** the experiment **twice more. Record** the results **of all three trials in Table 10-1**.

4.   Calculate the **average balloon diameter** for your three trials of exhaled air **at rest. Record** the average in **Table 10-1**.

5.   a. Take **one very deep breath** and exhale as much air as possible into the balloon.

       Inhale as much air as you can and **exhale as much air as you can** into the balloon. **Clamp and measure** the balloon as in step 3 above.

b. Record your results in **Table 10-1**.

6. Calculate the **average balloon diameter** for your three trials of exhaled air **while deep breathing. Record** the average in **Table 10-1**.

7. **Use the graph** in **Figure 10-8** to convert the average diameter of the balloon filled by **a normal breath** and the balloon filled by **a deep breath** into **cubic centimeters (cc) of lung volume.**

   Record the information in **Table 10-1**.

TABLE 10-1	
MEASURING LUNG CAPACITY	
DIAMETER (CM) OF BALLOON FILLED BY NORMAL BREATH:	
Trial 1	
Trial 2	
Trial 3	
Average	
Volume of air (cc) in balloon filled with normal breath:	
DIAMETER (cm) OF BALLOON FILLED BY DEEP BREATH	
Trial 1	
Trial 2	
Trial 3	
Average	
Volume of air (cc) in balloon filled with deep breath:	

8.   What is your **tidal volume?** _____

     What is your **vital capacity?** _____

Note:

Your measurement of tidal volume may be slightly higher than your actual tidal volume, since you also exhaled the reserve air held in your lungs.

---

9.   Which volume of air is exchanged while you are sleeping? _____

     Which volume of air is exchanged while you are running? _____

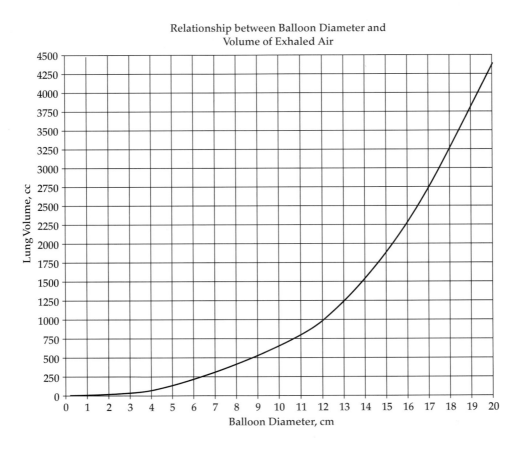

FIGURE 10-8.   Relationship Between Balloon Diameter and Volume of Exhaled Air

Check your answers with your instructor before you continue.

Self Test

Math the definitions in the right column with the terms in the left column. Each answer can be used only once.

1. _____ Anterior   a. Back

2. _____ Dorsal   b. Tail end of the body

3. _____ Posterior   c. Head end of the body

4. _____ Ventral   d. Belly side

5. Your lab partner arrived late to class on the day you started your fetal pig dissection. You have already selected a male pig for your dissection and you want your partner to choose a female. **Describe** how your partner can recognize and select a female pig from the container. **Be specific.**

6. Erin is playing shortstop in her college baseball team when she is hit in the throat by a line drive. On the way to the hospital, she has a very difficult time breathing. Which of the following is the most likely cause of her problem? **Explain** your answer.

   a. Bruised diaphragm   d. Crushed trachea
   b. Concussion   e. Blocked esophagus
   c. Broken rib cage

7.   Name **one part of the digestive system that** is located in the **thoracic** cavity.

8.   Name a structure that lies **between** the thoracic and abdominal cavities.

9.   When breathing, you are never able to completely fill your lungs with fresh air. A certain volume of stale air (with the oxygen removed) always remains in the air passageways. In some lung diseases, such as **emphysema,** large amounts of stale air accumulate and cannot be expelled. The lungs are filled with air that is useless for gas exchange. **Vital capacity** is significantly less than in normal lungs.

     Predict what will happen when a person with emphysema exercises heavily. **Explain** your answer.

10. In regard to your answer to question 9, explain how lack of air affects the body tissues. In your answer, use the following terms: **energy, oxygen, mitochondria, ATP, and red blood cells.**

Organs of the Abdominal Cavity

Objectives

After completing this exercise, you should be able to:

- identify and explain the function of each of the major structures of the abdominal cavity
- explain how the stomach and small intestine are specialized to perform specific functions
- compare and contrast pig anatomical features with human anatomical features
- apply your knowledge of the benefits of increasing surface area in the radish root to similar modifications in the intestine and lung
- relate your observations of earthworm locomotion to peristaltic action in organs of the digestive tract.

ACTIVITY 1   DISSECTION OF THE ABDOMINAL CAVITY

1. **Get your pig** from the storage container and set it up on a tray as you did in the last laboratory period.

    Get a set of dissecting instruments.

2. Use your **scissors** to cut through the **skin and muscles** in the abdominal region.

    Pull up on the umbilical cord to hold the skin away from the abdominal organs and cut from **number 5 to number 7 (see Figure 11-1).**

    Make two more incisions **around the umbilical cord (number 7 to number 8).**

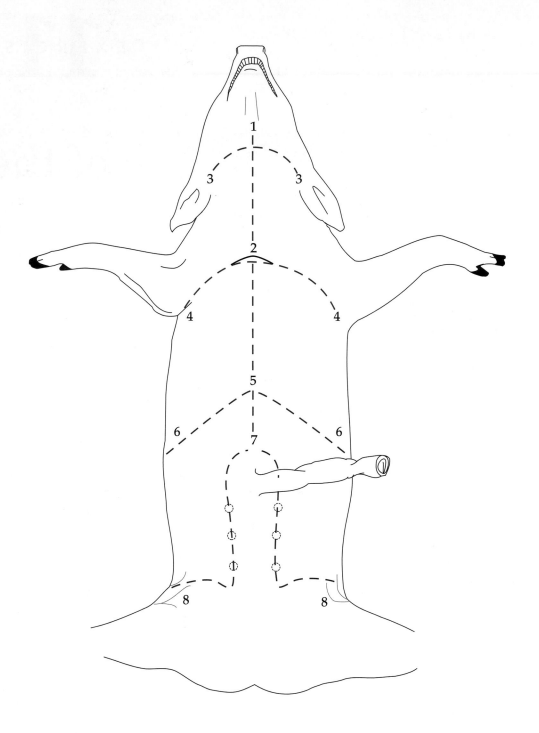

FIGURE 11-1.   Pattern for Body Cavity Incisions

3. The flap of tissue created by these cuts contains the **umbilical vein, the two umbilical arteries,** and the **urinary bladder.**

   Cut the umbilical vein **close to the umbilical cord.** This will allow you to turn back this flap so that it lies between the hind legs.

Hint:

If needed, rinse out the entire body cavity and blot it dry with paper towels. This will make your dissection experience much more enjoyable.

---

4. Just as in the thoracic cavity, the abdominal cavity is sealed with a smooth membrane called the **peritoneum.**

5. The largest organ in the abdominal cavity is the **liver.** It is reddish-brown with several lobes and is located just **posterior to the diaphragm.**

   The liver has many functions including **storage of energy reserves, detoxification of poisons, and bile formation.**

6. And speaking of bile, lift the **right lobe** of the liver and locate the **gall bladder,** a small saclike organ.

   The gall bladder **stores bile produced by the liver. Bile aids in fat digestion.** The bile is released through the **bile duct** into the small intestine, where fat digestion takes place.

7. Raise the liver to locate the **stomach,** a large, hollow bag located on the left side of the body.

   The stomach stores food and releases it gradually into the small intestine. Digestion of proteins begins here.

8. The long, dark red organ lying to the left of the stomach is the **spleen.**

   The spleen **functions as part of the immune system** and also **removes damaged and worn-out blood cells from circulation.**

   Within the abdominal cavity, the organs are held in position by membranes called **mesenteries.** You can see this membrane attaching the spleen to the stomach.

9. The **pancreas** is located between the stomach and the small intestine. To expose this organ, lift the stomach and use your **blunt probe** to dissect through the membranes in this area. The pancreas is an **elongated, pale, granular** organ positioned **across the body cavity.**

   The pancreas **secretes digestive enzymes** through a duct into the small intestine. It also functions as an **endocrine organ,** producing **two hormones that control blood sugar level.**

10. The **small intestine,** on the right side of the abdominal cavity, is a long, thin tube. Most chemical digestion occurs here. In the small intestine, **enzymes break down proteins, fats, carbohydrates, and nucleic acids.**

    Hold up one of the loops of the small intestine and notice that it is held together by **mesenteries.** If you look closely, you can see **arteries and veins** fanning out across the membrane.

    Mesenteries provide a **support structure** for blood vessels and nerves leading to all the abdominal organs.

11. The **large intestine** is on the left side of the abdominal cavity. The main function of the large intestine is the **reabsorption of water from wastes,** preventing dehydration. The concentrated wastes are stored in the **rectum** and eliminated through the **anus.**

12. Lift the intestines on either side of the body and you will see a **kidney.** The kidneys **remove nitrogen-containing wastes and form urine.**

13. The **urinary bladder,** located between the two umbilical arteries, is a **temporary storage organ for urine.** When the bladder fills, nerve endings in its muscular walls are stimulated, causing the muscles to contract for urination.

14. **Label** the following structures on **Figure 11-2: liver, gall bladder, stomach, spleen, pancreas, small intestine, large intestine, mesenteries, kidney, and rectum.**

Check your figure labels with your instructor before you continue.

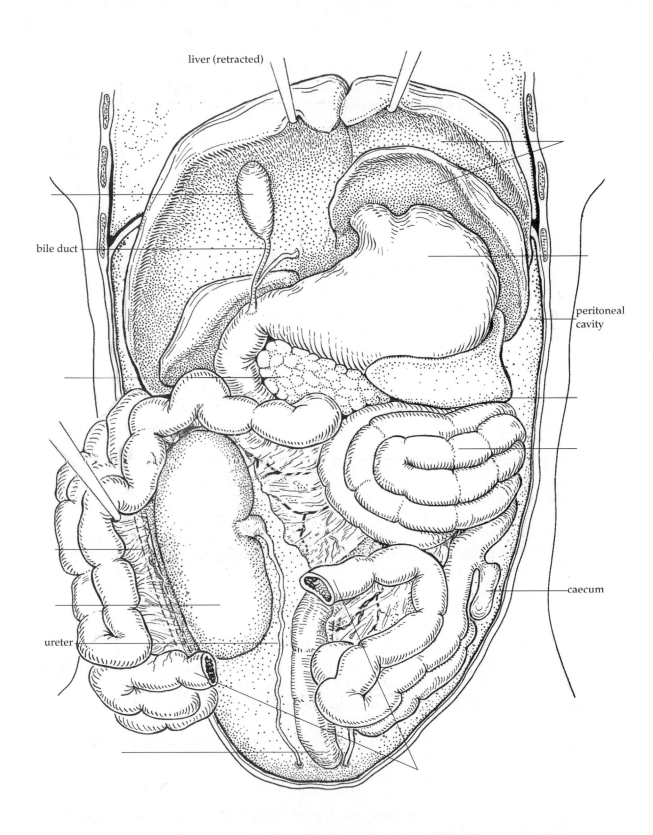

liver (retracted)

bile duct

peritoneal cavity

caecum

ureter

FIGURE 11-2. Organs of the Abdominal Cavity

ACTIVITY 2   THE URINARY SYSTEM

Preparation

Waste products are removed from the body by several systems. For example, **carbon dioxide,** a waste product of **cell respiration,** is removed by the **lungs. The nitrogen-containing** waste products of **protein digestion** are excreted by the urinary system.

The urinary system includes the two **kidneys,** which **produce urine;** the two **ureters,** which **carry the urine to the bladder;** the **bladder,** which **stores the urine;** and the **urethra,** which **transports urine** out of the body through the **urogenital or urinary opening** (depending on your sex).

The kidneys also have an important function in maintaining **homeostasis** (keeping internal body conditions stable). The body must maintain adequate levels of salts, sugars, and other substances in the bloodstream, while disposing of the excess. The kidneys play a key role in this process, keeping **pH** values, **fluid and salt content,** and **plasma levels of valuable materials** nearly constant.

1. Push the intestinal mass to the **left side** of the abdominal cavity, exposing the **right kidney.** The kidneys are located along the **dorsal wall of the abdominal cavity** (see **Figure 11-3**).

2. You will notice that the kidney is covered by a thick layer of connective tissue, the **peritoneum.**

    Using the blunt probe, carefully remove the peritoneum, uncovering the right kidney and ureter.

    In addition to the ureter, you will see the **renal artery and vein.** The renal **artery** carries blood **to the kidney for filtration.** The blood **returns** to the body circulation through the **renal vein.**

3. Each kidney is drained by a **ureter,** a tube leading from the kidney to the urinary bladder.

    Locate the ureter and follow it to the **bladder,** which is located in the flap of tissue created by your **incisions around the umbilical cord in Activity 1.** The bladder can be found **between the two umbilical arteries.**

    The **posterior end of the bladder** narrows to form a tube, the **urethra,** which transports urine out of the body. Most of the urethra cannot be seen, because it is positioned deeper into the pelvic cavity.

4. Label the following structures on **Figure 11-3: kidney, urethra, urogenital opening,** and **bladder.**

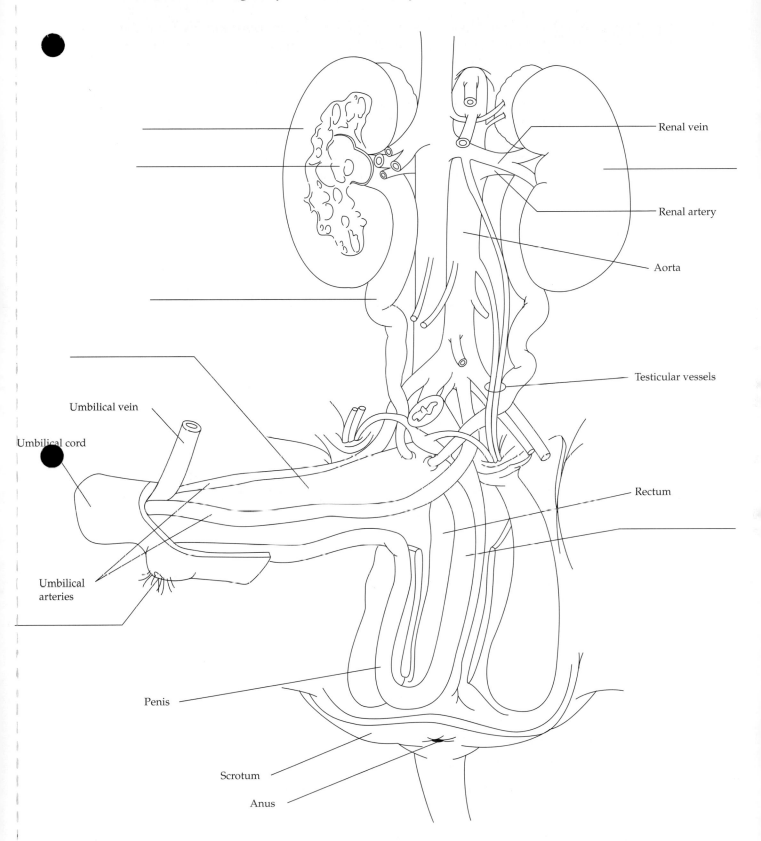

FIGURE 11-3. Urinary System of the Male Pig

5. Cut the ureter and renal blood vessels. Remove the intact right kidney from the abdominal cavity. If necessary, remove additional peritoneum so that the kidney can be removed without damage.

6. Make an incision through the long axis of the kidney, cutting it into two equal halves (like opening a book).

    Within the kidney, you will see an **outer area** of densely packed tissue, the **cortex.** Urine formation begins in microscopic structures within this region.

    The **inner** section, the **medulla,** appears fibrous. This area collects the urine and funnels it to the ureter.

    Observe the connection of the ureter to the kidney near the center of this region.

7. Label the following structures on the dissected kidney in **Figure 11-3: cortex, medulla,** and **ureter.**

Comprehension Check

Fill in the blanks with the choice that is most appropriate to describe the function of each part of the urinary system. Answers can be used **more than once.** Some questions may have **more than one correct answer.**

a.	Ureter	d.	Renal artery	g.	Urinary bladder
b.	Kidney	e.	Renal vein	h.	Medulla
c.	Urethra	f.	Peritoneum	i.	Cortex

1. _____ This membrane holds the kidneys in position.

2. _____ If a drop of urine is in the ureter, which **TWO** structures will it pass through next (**in sequence**)?

3. _____ Carries blood high in metabolic wastes to the kidney.

4. _____ Returns filtered blood to body circulation.

5. _____ When this structure is full, nerve endings signal the urge to urinate.

6. _____ Urine formation begins here.

7. _____ Either urine or semen could exit the body through this tube.

8. _____ Plays a vital role in maintaining homeostasis in the body.

9. _____ This portion of the kidney funnels urine to the ureter.

Check your answers and Figure 11-3 labels with your instructor before you continue.

ACTIVITY 3   INTERNAL STRUCTURE OF THE STOMACH

1. With your fingers, locate the **pyloric sphincter** of the fetal pig. It will feel like a **small, hard mass** in the stomach wall. Using the same method, locate the **cardiac (gastroesophageal) sphincter**.

    Referring to **Figure 11-4**, use your **scissors** to remove the entire stomach, being careful to include **both sphincters**.

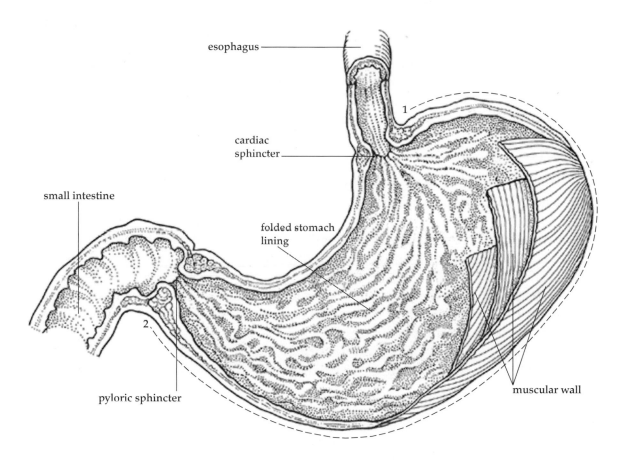

esophagus

cardiac sphincter

small intestine

folded stomach lining

pyloric sphincter

muscular wall

1

2

FIGURE 11-4.   Internal Structure of the Stomach

2. Following the **dotted line** in **Figure 11-4**, cut from **number 1 to number 2** and open up the stomach.

    Rinse the inside of the stomach and **blot it dry** with paper towels.

3. Examine the **interior** of the stomach. **Observe** that the stomach wall is folded, like an accordion.

    Imagine that, disregarding what you know about dietary guidelines, you stuff yourself with two cheeseburgers, a supersize order of fries, an apple pie, and a chocolate shake.

    During this intake, what happens to the **size** of your stomach? _____

    How are the **folds in the stomach wall** related to this change in size?

4. A sphincter is a **ring of muscle** that surrounds the opening of a hollow organ. When the sphincter muscles **contract**, the opening is shut.

    Visualize what happens when you pull the ties on a plastic trash bag and the bag closes. This is similar to the action of a sphincter muscle.

    The stomach has **two sphincters**. One is between the esophagus and the stomach (the **cardiac sphincter**) and the other is between the stomach and the small intestine (the **pyloric sphincter**).

    (Circle one answer.) If there is food in the esophagus, the cardiac sphincter will be **opened/closed**.

    What is happening to the food in the **stomach** while **both sphincters are closed**?

    If the **pyloric sphincter is open**, where is the food going? _____

5. With one hand, feel the pyloric sphincter. At the **same time**, use the other hand to feel the cardiac sphincter.

    How are the two sphincters **different**?

    Which sphincter do you think is **stronger**? _____

    The **cardiac sphincter** prevents food in the stomach from backing up into the esophagus.

    The **pyloric sphincter** controls the **rate** at which food enters the small intestine. No matter how full your stomach is, food is always released **gradually** into the small intestine.

Comprehension Check

1. Considering your examination of the stomach, **explain** what happens during **vomiting**.

2. What would happen if your **pyloric sphincter** were not able to **close**?

3. What would happen if your **pyloric sphincter** were not able to **open**?

4. What would happen if the **cardiac sphincter** didn't close completely and stomach juices were allowed to enter the esophagus?

ACTIVITY 4　　　　　　　　　EXAMINING THE SMALL INTESTINE

1. Hold up a loop of the small intestine and **examine the mesenteries**.

    In addition to holding the small intestine in position, the mesenteries provide a passageway for **large numbers of blood vessels**.

2. Using your **scissors**, carefully **snip through the mesenteries** and start separating the loops of the small intestine.

    Remove the small intestine.

3. Lay the intestine out on the laboratory counter and use a **meter stick or yardstick** to measure the length.

    The length of my pig's small intestine is _____.

    (Circle one answer.) The small intestine is **shorter/longer** than I expected.

    If the small intestine is longer than you expected, you might be interested in the origin of the name "small" intestine. The intestine is named for its **small diameter (and not its length)**.

    The large intestine has twice the diameter of the small intestine **(but only half the length)**.

4. Although each person is slightly different, the human small intestine averages about **20 feet (6 meters)** in length.

    Using your knowledge of the digestive system, **explain** how increasing the length of the small intestine would help this organ **perform its functions**.

5. Remove approximately **two inches (5 cm)** from the small intestine. Using your **scissors**, cut **open** the piece of intestine **lengthwise** and place it on the stage of the **dissecting microscope**.

6. Observe that the inside of the small intestine is **not smooth**.

    The interior wall is **folded**, and **each fold is carpeted** with slender fingerlike projections called **villi** (singular is **villus**).

7. Although you cannot see them at this magnification, each of the villi is covered, in turn, with even more tiny projections called **microvilli** (see **Figure 11-5**).

    Located inside each of the villi is a network of capillaries and lymphatic vessels (**lacteals**). These vessels pick up nutrients as they are absorbed, so that they can be distributed to all parts of the body.

    The small intestine is quite a narrow tube. However, if the villi and the microvilli in the intestinal tract were laid out flat, the surface area would be **half the size of a basketball court**!

8. After you have completed your dissection, do the following:

    a. Remove the strings from the limbs.
    b. Your instructor will tell you how to dispose of your pig and any remaining tissues.
    c. Wash and dry all instruments and trays.

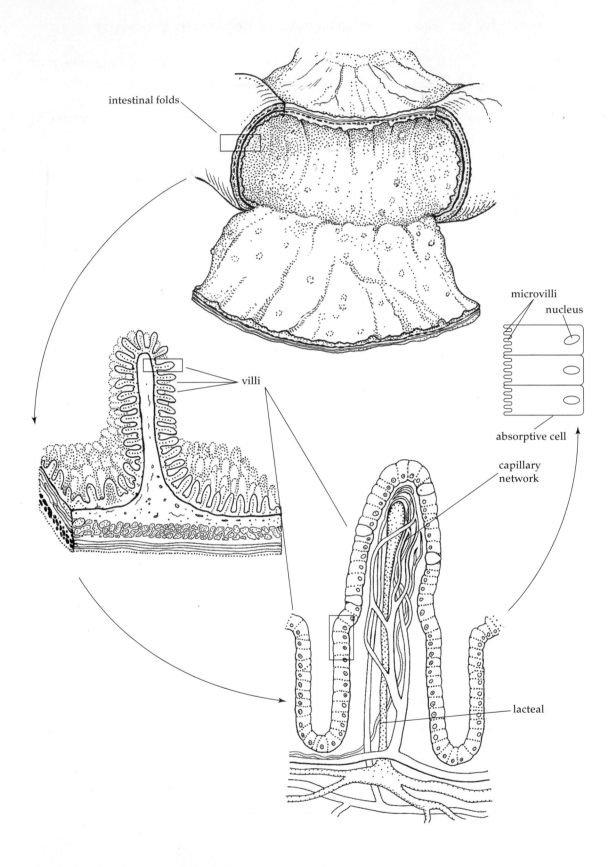

FIGURE 11-5. Interior of the Small Intestine

ACTIVITY 5   THE IMPORTANCE OF SURFACE AREA

1. **Why does the small intestine need such a large surface area?** The need to increase surface area is common throughout the plant and animal kingdoms. On the laboratory counter, you will find a demonstration of one method used to increase surface area in a plant root (**a radish seedling**).

   (Circle one answer.) The radish root tip appears **smooth/fuzzy**.

   What **two** important substances enter a plant through the **roots?**

   _____   _____

2. The many slender extensions you can see covering the root of the radish seedling are called **root hairs**.

   Use **Figure 11-6** to determine how **root hairs** help plant roots function.

   In **both** the pictures in **Figure 11-6, the dark line represents the outside of the root surface.**

   Following the dark line, **place an X in every square that the line passes through** (any square that touches the exterior of the root surface). A few Xs have been drawn into the figure to get you started.

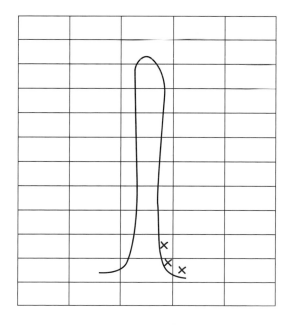

Without Root Hairs

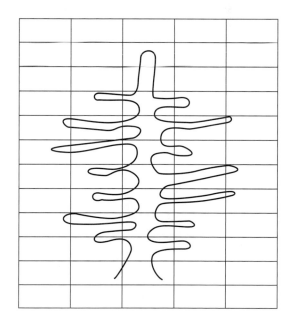

With Root Hairs

FIGURE 11-6. Comparison of Surface Area with and without Root Hairs

3. For each picture in **Figure 11-6, count the number of squares with an X** and enter your results below:

    Without root hairs _____ **With root hairs** _____

4. Imagine that each graph square represents **one drop of water**. Which root (with or without root hairs) can absorb the **greatest number of water droplets at one time? Explain** your answer.

Comprehension Check

1. Imagine that each graph square represents **one nutrient molecule**.

   Which type of intestinal lining (with or without villi) can absorb the **greatest number of nutrient molecules at one time? Explain your answer.**

2. Considering your answer to the above question, how are **microvilli** beneficial for small intestine function?

Check your answers with your instructor before you continue.

ACTIVITY 6

A CLOSER LOOK AT THE INSIDE
OF THE SMALL INTESTINE

1. Referring to **Figure 11-7**, locate the **villi** (projecting toward the center of the tube).

2. At the base of the villi, you will see many **small**, circular structures.

   These are the **intestinal glands**. Intestinal glands produce **digestive enzymes** that perform the final breakdown of food molecules.

   Add a **label** for **intestinal glands** to the diagram in **Figure 11-7**.

3. Opposite the intestinal glands, you will see a cluster of **larger** circles.

   These are **lymphoid nodules,** small lymph nodes that act as **bacterial filters** in the digestive, respiratory, and urinary tracts.

   Which other body locations contain **lymph nodes**?

   Add a **label** for **lymphoid nodules** to the diagram in **Figure 11-7**.

4. You can see **two layers of muscle tissue** around the outside of the intestine.

   The **innermost** layer of muscle fibers runs in a circle around the tube and is referred to as **circular muscle**.

   The **outermost** layer has the muscle fibers arranged along the **length** of the tube and is called **longitudinal muscle**.

   Add **labels** for the **circular and longitudinal muscle** layers to the diagram in **Figure 11-7**.

5. The **circular and longitudinal** muscle layers **alternately** contract, producing mixing movements and moving food through the small intestine with waves of **peristalsis**.

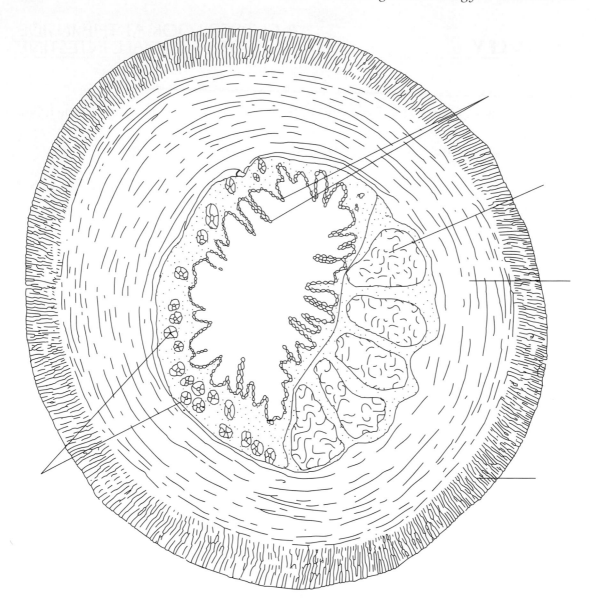

FIGURE 11-7. Cross Section of the Small Intestine

Check your figure labels with your instructor before you continue.

ACTIVITY 7   EARTHWORM LOCOMOTION

Preparation

To see waves of **peristaltic action,** we will examine locomotion in an earthworm.

Earthworm movement is a result of alternating muscle layers in the body wall. Alternating contractions of the **circular and longitudinal muscle** layers allow the earthworm to move forward. Segments may also move in the reverse direction, allowing the earthworm to move backward.

1. Work in groups. **Get an earthworm** and **several clean, dampened paper towels.**

Attention:
Wash your hands before touching the worms!

2. **Lay out** the dampened paper towels on your laboratory counter and **place the earthworm** on the towels.

    Observe the **muscle contractions** as the earthworm moves forward. From your observations, **describe how the earthworm's body changes** as it moves.

3. For a closer look at the alternating action of circular and longitudinal muscles, place the earthworm on the stage of your **dissecting microscope.** Similar muscle contractions in the small intestine mix and move food.

    When you have completed your observations, **return the earthworm** to the container on the supply table.

4. How is the **process of peristalsis** related to earthworm locomotion? (If you need more information to answer this question, refer to your textbook.)

5. **(Circle one answer.)** The circular and longitudinal muscles in the small intestine are composed of **smooth muscle tissue/skeletal muscle tissue/cardiac muscle tissue**.

Self Test

Fill in the blanks with the **most appropriate** answer. Answers can be used **only once**.

a. **Lymphoid nodules**   e. **Longitudinal muscle**
b. **Root hairs**   f. **Peristalsis**
c. **Villi**   g. **Surface area**
d. **Circular muscle**   h. **Mesenteries**

1. _____ Membranes that hold the small intestines in position.

2. _____ Small projections that increase surface area in the small intestine.

3. _____ Involuntary muscle running **around** the intestine.

4. _____ Contractions that move food through the digestive tract.

5. _____ The **outermost** muscle layer in the small intestine.

6. _____ Has a **function similar** to the villi of the small intestine.

7. Which of the following organ systems are represented in the **abdominal cavity?** **Circle ALL correct answers** and **give an example of an abdominal organ or structure** that belongs to each system.

ORGAN SYSTEM	EXAMPLES
a. Digestive system	
b. Respiratory system	
c. Urinary system	
d. Circulatory system	

8. When you experience **heartburn,** which **stomach** structure is not functioning correctly? **Explain your answer.**

9. **Bulimia** is an eating disorder that involves forced vomiting. Over a period of years, bulimics gradually lose the enamel coating from their teeth. Using your knowledge of the digestive system, explain what causes the loss of tooth enamel.

10. **Urinary incontinence** (inability to control urine flow) frequently occurs in the elderly. Which organ of the urinary system may be malfunctioning? Which structure is malfunctioning in that organ you named? **Explain your answer.**

11. **Figure 11-8** contains two designs for air sacs in a human lung. Which will be the **most efficient** in taking in oxygen and removing carbon dioxide? **Explain your answer**.

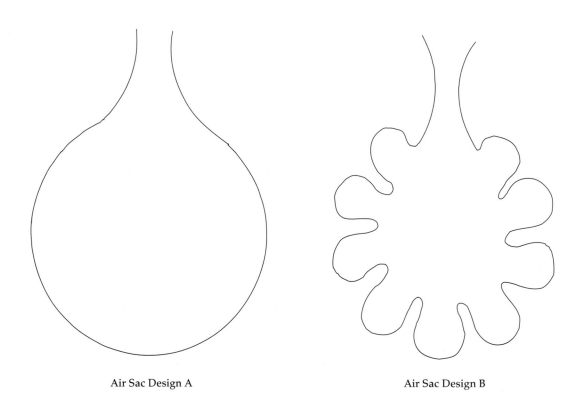

Air Sac Design A   Air Sac Design B

FIGURE 11-8. Comparison of Air Sac Designs

Complete the crossword puzzle in **Figure 11-9** to review the pig structures and functions.

ACROSS
1. Pair of blood vessels that carry oxygen-poor blood that is high in wastes from the fetus to the placenta; called the umbilical _____
2. Most chemical digestion occurs here; enzymes break down proteins, fats, carbohydrates, and nucleic acids
3. External genital structure in female pigs
4. Part of the immune system; site of T-cell development
5. Muscular tube that connects the mouth to the stomach
6. Part of the immune system; blood-filtering organ and storage site for red blood cells and some white blood cells
7. Thin connective tissue membranes that hold the internal organs in place and provide a passageway for blood vessels and nerves that supply the organs
8. Bony roof of the mouth
9. Houses the vocal cords, the vibrations of which make speech possible; composed of cartilage
10. Fleshy sac that contains the testes in males
11. Endocrine (hormone-producing) gland; produces a hormone that helps in regulating the metabolic activity of the body
12. Secretes digestive enzymes into the small intestine; also functions as an endocrine organ, producing two hormones that control blood sugar levels
13. Has many functions including storage of energy reserves, detoxification of poisons, and bile formation

DOWN
1. Stores bile formed by the liver; releases it through the bile duct into the small intestine (for fat digestion)
2. Baglike organ that stores food and releases it gradually into the small intestine; mechanical and some chemical digestion occurs here
3. Air channel to the lungs; surrounded by tough, elastic rings of cartilage
4. Excretory organ that stores urine
5. Hood-shaped flap of tissue that covers the opening to the trachea; prevents food and liquid from entering the air passageways
6. Composed of millions of tiny sacs, which are the site of gas exchange for the whole body; oxygen enters and carbon dioxide is removed
7. Paired blood vessels that drain deoxygenated blood from the head region back to the heart; called the jugular _____
8. Excretory organ that removes nitrogen-containing wastes and maintains the proper solute concentration of body fluids
9. Digestive organ whose main function is the reabsorption of water from wastes; prevents dehydration
10. Muscular pump consisting of four chambers; pumps blood to the lungs and the body tissues

FIGURE 11-9. Crossword Puzzle

The Circulatory System

Objectives

After completing this exercise, you should be able to:

- diagram the pattern of blood circulation around the body and label the blood in each location as oxygenated or deoxygenated
- differentiate between the pulmonary and systemic circuits
- identify and explain the function of each of the major structures of the heart
- discuss the connection between blood vessel diameter and the rate of blood flow
- discuss the effect of exercise on pulse rate, blood pressure, and the efficiency of oxygen transport
- discuss the ways in which exercise changes the distribution of blood in the body
- apply your knowledge of vessel and heart structure to cardiovascular health issues.

CONTENT FOCUS

We hear a lot about exercising our muscles and keeping fit. Activities that increase cardiovascular fitness are an important part of any exercise program. Many people don't realize that the heart is a muscular pump, and so, as with all our body muscles, heart function can be improved by exercise and activity.

The average heart is only about the size of your fist. It normally beats about **72 times per minute.** This may not sound too impressive, but each day, the heart pumps **1500 gallons** of blood. Over your lifetime, this adds up to enough blood to fill **13 supertankers!**

Each beat of the heart **transports blood** that contains **food, oxygen,** and **hormones** to all the cells of the body and **removes wastes** such as **carbon dioxide.**

The circulatory system of humans and most other animals consists of **three basic elements: a pump, blood vessels, and blood (the circulatory fluid).** In this exercise, you will take a closer look at the heart and do some experiments to demonstrate how the circulatory system adjusts to the needs of the body.

ACTIVITY 1

Preparation

The heart is divided into **four** chambers. The two **upper** chambers, called **atria**, receive blood returning to the heart. The two **lower** chambers, called **ventricles,** pump blood out.

1. The circulatory pathway that carries blood through the lungs and back to the heart is called the **pulmonary circulation.** The circulatory pathway that carries oxygenated blood through the upper and lower body and back to the heart is called the **systemic circulation.** Following these pathways, you can see that **blood returns to the heart TWICE as it circulates around the body.**

   As shown in **Figure 12-1, blood returns from the body** into the **right atrium.** The blood then travels into the **right ventricle,** which pumps it out to the **lungs.**

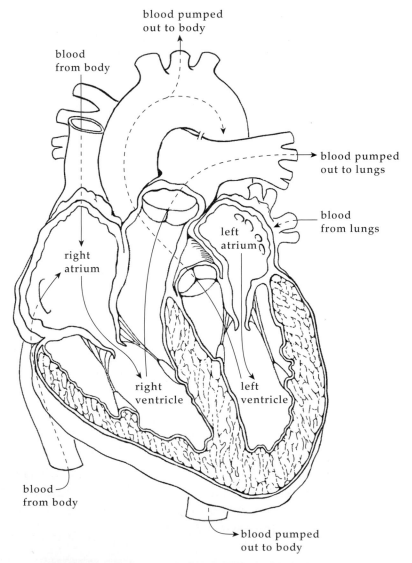

FIGURE 12-1. Circulation of Blood Around the Body

Blood **returns from the lungs** into the **left atrium,** then goes into the **left ventricle,** which pumps it to the **rest of the body.**

2. Using **a colored pencil or highlighter** (and the information you studied about lung function), color in **blue** each **heart chamber in Figure 12-1** that contains **deoxygenated** blood (blood **low** in oxygen). Color in **red** each **heart chamber** that contains **oxygenated** blood (blood **high** in oxygen).

3. Within the pulmonary and systemic circulatory systems, the blood travels through **arteries, veins,** and **capillaries.**

   Arteries always transport blood **away from the heart. Veins** carry blood **toward the heart. Between the arteries and veins** lie **beds of capillaries** where cells pick up oxygen and release carbon dioxide.

   On **Figure 12-2, place a letter "A" next to** each blood vessel that is an **artery.** Place a **letter "V" next to** each **vein.** Place a **letter "C"** next to each **capillary bed.**

4. On **Figure 12-2,** color in **blue** each **heart chamber and blood vessel** that contains **deoxygenated** blood.

   Color each **heart chamber and blood vessel** that contains **oxygenated** blood in **red.**

5. Using the **clues** below, correctly **label** the listed blood vessels on **Figure 12-2.**

BLOOD VESSEL	CLUE FOR IDENTIFICATION
Superior vena cava	Vessel that returns blood from the head to the heart
Inferior vena cava	Vessel that returns blood from the lower body to the heart
Pulmonary artery	Vessel that carries blood to the lungs
Pulmonary vein	Vessel that returns blood from the lungs to the heart
Aorta	Vessel that distributes oxygenated blood around the body
Carotid artery	Branch off the aorta that delivers blood to the head

✔ Comprehension Check

1. **(Circle one answer.)** A drop of blood in the **pulmonary artery** is **oxygenated/deoxygenated.**

   Which heart chamber did it come from? _____

   Where will this drop of blood be found next? _____

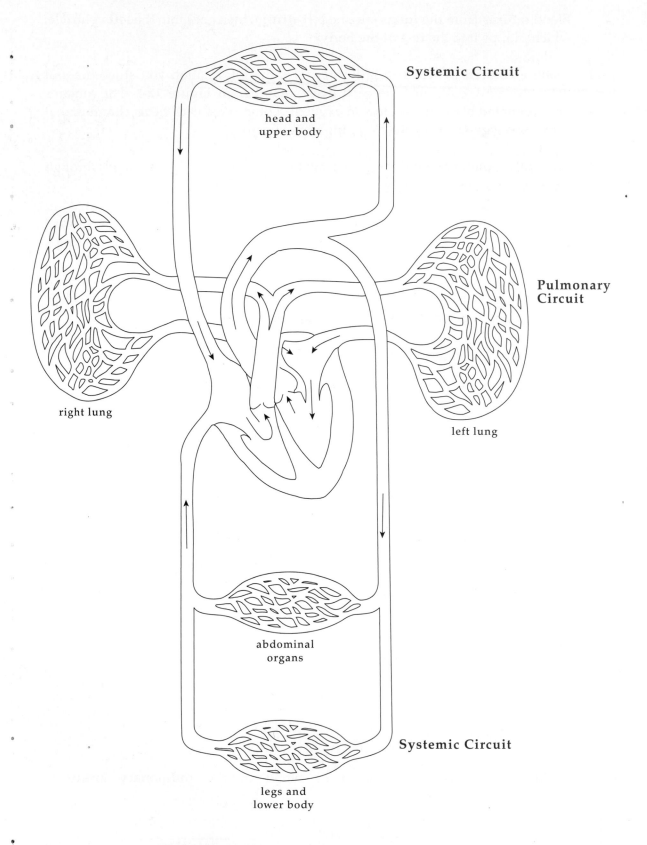

Systemic Circuit

head and
upper body

**Pulmonary
Circuit**

right lung

left lung

abdominal
organs

Systemic Circuit

legs and
lower body

FIGURE 12-2. Pulmonary and Systemic Circulation

2. **(Circle one answer.)** The blood flowing through the **left side** of the heart is **oxygenated/deoxygenated. Explain your answer**.

3. **(Circle one answer.)** A drop of blood in the **inferior vena cava is oxygenated/deoxygenated.**

   Which heart chamber will it enter next? _____

   Where will this drop of blood travel next? _____

4. Do all arteries carry **oxygenated blood? Explain your answer**.

Check your answers and Figure 12-2 labels with your instructor before you continue.

ACTIVITY 2   A CLOSER LOOK AT THE HEART

1. Work in groups. Get a **model of the human heart.** Using **Figure 12-3** as a guide, locate the **four chambers** on the **heart model. Observe** the muscular walls of the left and right ventricles on the **heart model.**

   On which side of the heart is the muscular wall the **thickest?** _____
   Why?

Hint:
Review the information in Figures 12-2 and 12-3.

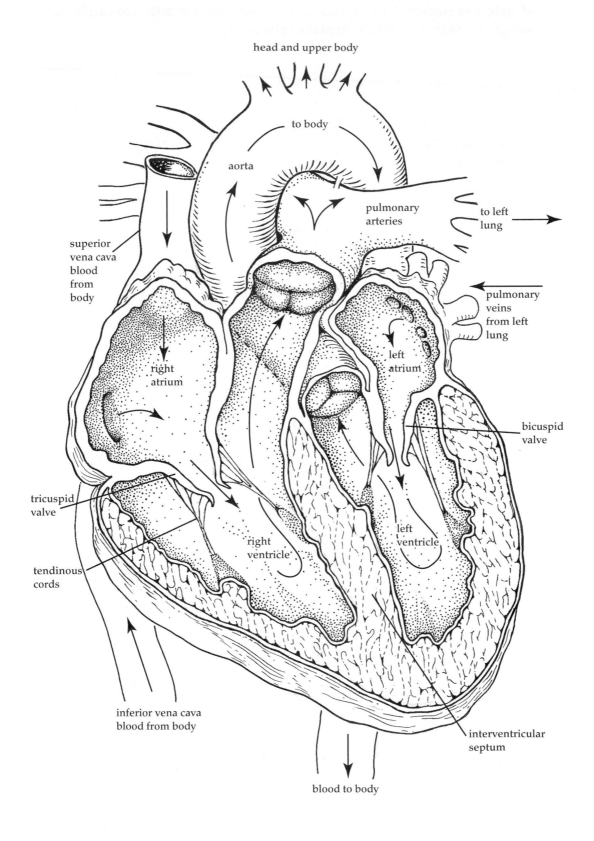

head and upper body

to body

aorta

pulmonary arteries

to left lung

superior vena cava blood from body

pulmonary veins from left lung

right atrium

left atrium

bicuspid valve

tricuspid valve

tendinous cords

right ventricle

left ventricle

interventricular septum

inferior vena cava blood from body

blood to body

FIGURE 12-3. Interior View of the Heart

2. **Heart valves** prevent blood from flowing **backward** through the heart. Each valve section is anchored to the heart wall by **tendinous cords.** These strong cords hold the valve sections **closed** when the ventricles contract.

   On your model of the heart, locate **one valve between the right atrium and the right ventricle** (the **tricuspid valve**) and a **second valve** between the **left atrium and the left ventricle** (the **bicuspid** or **mitral valve**).

3. To beat properly, the muscles of the heart must get an adequate blood flow.

   Look at the **outer surface** of the heart model. You will see many **small arteries and veins** branching across the surface of the heart.

   The blood that flows through the heart is moving too fast to be used as a source of oxygen and nutrients by cardiac muscle cells.

   A branch off the **aorta** connects to the **coronary arteries,** which leads to capillary beds in the heart muscle. Blood is returned through the **coronary veins** to the inferior vena cava.

✔ Comprehension Check

1. **Circle the correct answer** in each of the following statements.

   a. The **bicuspid** valve is **opened/closed** when the **left atrium** contracts.

   b. The **tricuspid** valve is **opened/closed** when the **right ventricle** contracts.

   c. The **tendinous cords** of the **tricuspid** valve are **tightened/relaxed** when the right ventricle contracts.

2. Blood that leaves each ventricle is pumped into an artery. Another valve prevents blood from flowing backward into the ventricle. These are called the **pulmonary and the aortic semilunar valves. Locate** the semilunar valves on the heart model (see **Figure 12-4**).

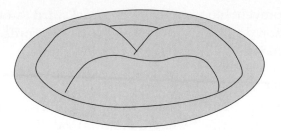

FIGURE 12-4. Semilunar Valve

3. The semilunar valve on the **right** side of the heart is between **which heart chamber and which blood vessel?**

   _____   _____

4. The semilunar valve on the **left** side of the heart is between **which heart chamber and which blood vessel?**

   _____   _____

5. **(Circle one answer.)**

   a.   When the right **ventricle** is **contracting,** the **pulmonary** semilunar valve is **opened/closed.**

   b.   If a drop of blood is going **toward the lungs,** it has passed through the **pulmonary/aortic** semilunar valve.

6. *Challenge Question!* When the right **atrium** is **contracting,** the **pulmonary** semilunar valve is **opened/closed.**

Check your answers with your instructor before you continue.

ACTIVITY 3

Preparation

The ventricles contract to pump blood through all the arteries of the body. As blood travels away from the heart, it passes through smaller and smaller blood vessels.

How hard does the heart have to work to pump blood through blood vessels of different sizes?

How much force the heart exerts to pump blood **depends on how easily fluid passes through the blood vessel.** Using **rubber tubing to simulate blood vessels,** you will design an experiment to determine whether the **diameter of blood vessels** affects the **rate of blood flow.**

During your experiment, you want to consider **only** the effect of blood vessel diameter. For this reason, you will **not** use a pump to simulate heart action. Blood will flow through your simulated blood vessels using only the force of **gravity.**

1. Work in groups.

   Get the following supplies: **one large container with spigot; one stopwatch; one plastic bucket; one large graduated cylinder;** and **three pieces of plastic tubing** (see sizes below):

Large	5/16" inside diameter
Medium	4/16" inside diameter
Small	3/16" inside diameter

2. **Develop a hypotheses** about the relationship between the **blood vessel diameter** and the **rate of blood flow.** Write it in the space below.

BLOOD FLOW HYPOTHESES

Check your hypotheses with your instructor before you continue.

3. Carefully **develop a plan** about how you will get the information you need to test your hypotheses. **List all the steps** of your plan, **including** the **equipment** you will use.

STEPS OF YOUR EXPERIMENTAL PLAN

4. In the space below, **make a table or chart** that shows your results clearly and neatly. Collect your experimental data and **record the results**.

5. **Plot a graph of your results in Figure 12-5.**

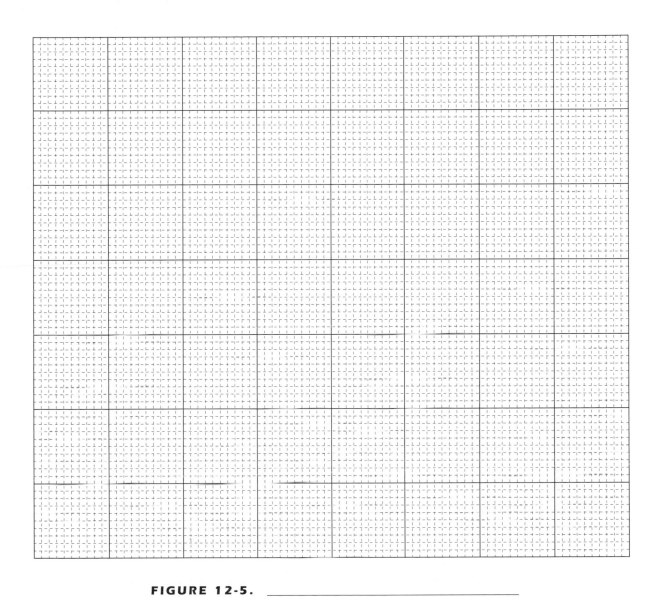

FIGURE 12-5. _____

6. Discuss the results with your group members. Write a couple of sentences that **summarize** your **results**.

Hint:

Remember—results include only facts, never opinions.

---

7. **Write a conclusion based on your hypotheses and collected data. Support your conclusion** by mentioning facts collected during your experiment.

8. **(Circle one answer.)** As blood travels from arteries into arterioles, the rate of blood flow will **increase/decrease**.

9. Capillaries are the **smallest** blood vessels. How does this fact affect the **rate** of blood flow?

10. Why is **flow rate** an important factor for determining how **efficiently** cells can pick up oxygen and remove carbon dioxide?

11. In the disease **atherosclerosis**, fatty plaques accumulate on the inside of artery walls. What effect would this have on the **rate** of blood flow? **Explain your answer.**

Check your answers with your instructor before you continue.

ACTIVITY 4 THE EFFECT OF EXERCISE

Preparation

Every time the ventricles contract **(systole)**, forcing blood out of the heart, pressure in the blood vessels **increases**. When the ventricles **relax (diastole)**, pressure in the blood vessels **drops** again. The pressure exerted by the blood on the walls of the blood vessels is called **blood pressure**. The original instrument used to measure blood pressure forced a **column of mercury** upward in a glass tube. Newer instruments that measure blood pressure still use the same measurements **(millimeters of mercury or mm Hg)**. The average blood pressure in a young adult male is approximately **120 mm Hg during systole, and 80 mm Hg during diastole**. This is expressed as **120/80**. Young adult females average about 8–10 mm Hg less than males. Blood pressure can be lower in adults that exercise regularly.

Blood is forced into your arteries during each contraction of the heart. You can feel this wave of blood moving through the arteries as a **pulse** in the carotid artery of your neck. The pulse can also be felt in the radial artery of your wrist and other locations in the body. The **number of pulses** tells you **how fast the heart is beating**. The average pulse rate in a young adult is **72 beats per minute**.

1. Working with a **group**, do the following experiments.

    One student will be the test subject, another will monitor the blood pressure, another will monitor the pulse rate, and the fourth will be the timekeeper and record the experimental data.

Caution!
DO NOT be the test subject for this activity if you have heart or blood pressure problems!

2. Get the following supplies: **one blood pressure monitor, a stopwatch, and an aerobic step**.

Caution!
Blood pressure monitors are fragile and expensive! Please be careful with this equipment.

3. Sit quietly for **one minute. While sitting**, take your **resting pulse rate and blood pressure**.

    Resting pulse rate: _____

    Resting blood pressure: _____

4. **Rapidly step up and down** on the aerobic step for **five minutes.** You should really be working out!

   Immediately sit down and record the **pulse rate and blood pressure.**

   Exercise pulse rate: _____

   Exercise blood pressure: _____

5. Is there a difference in pulse rate before and after exercise? _____

   If so, **describe the difference.**

6. Is there a difference in blood pressure before and after exercise? _____

   If so, **describe the difference.**

7. What is the heart doing when **blood pressure** increases? _____

   What is the heart doing when the **pulse rate** increases? _____

8. How do these **two different changes in heart action** help the body during exercise? In your answer, use the following terms: **blood flow, oxygen, carbon dioxide, cell respiration,** and **ATP.**

Self Test

1. Follow a drop of blood around the body. Begin with the tissue capillaries supplying your left big toe. **Place these locations in the correct sequence.**

   a. _1_ Tissue capillaries in big toe

   b. ___ Pulmonary artery

   c. ___ Left atrium

   d. ___ Pulmonary vein

   e. ___ Right atrium

   f. ___ Left ventricle

   g. ___ Inferior vena cava

   h. ___ Aorta

   i. ___ Right ventricle

   j. ___ Arterioles (small arteries)

   k. ___ Venules (small veins)

   l. ___ Lungs

2. If there were **no tendinous cords** attached to the tricuspid valve, and the right ventricle is contracted, where would the blood go?

3. If a person's pulse rate is 96 pulses per minute, what is this person's **heart rate**?

4. For the following body locations, enter **"D"** if the blood is **deoxygenated** and **"O"** if the blood is **oxygenated.**

    ___ Tissue capillaries entering big toe

    ___ Pulmonary artery

    ___ Left atrium

    ___ Pulmonary vein

    ___ Right atrium

    ___ Left ventricle

    ___ Inferior vena cava

    ___ Aorta

    ___ Right ventricle

    ___ Tissue capillaries leaving big toe

    ___ Arterioles (small arteries)

    ___ Venules (small veins)

5. You are a doctor listening to heart sounds through a stethoscope. While listening, you notice an unusual "hissing" sound in a patient's heart. On consulting the patient's medical records, you find that as a child he suffered from rheumatic fever, which causes scar tissue to form around the heart valves. What might be causing the hissing sound?

6. When a person **blushes,** blood flow to the skin increases. In order for this change in blood flow pattern to occur, how must the **diameter** of skin arteries change?

7. Place an **"X"** in front of the **one** measurement of oxygen concentration **that is most likely to be correct for all four locations.**

X	RIGHT VENTRICLE	PULMONARY ARTERY	PULMONARY VEIN	LEFT ATRIUM
	40	40	100	100
	100	40	40	100
	100	100	100	100
	40	40	40	40
	40	100	40	100

The graph in **Figure 12-6** shows the distribution of blood (by percentage) in the circulatory system when the body is at rest and during heavy exercise.

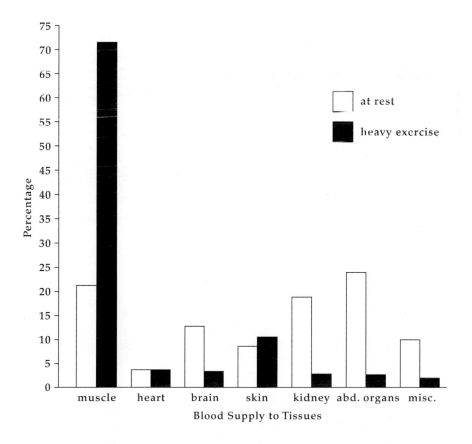

FIGURE 12-6. Distribution of Blood at Rest and During Exercise

8. In reference to the graph in **Figure 12-6**, what **change** occurs to the **percentage of blood supply to the muscles** when you are exercising?

    Considering the **change in the percentage of blood flow** to the muscles shown in the graph, what probably happens to the **diameter of the blood vessels** that supply blood to the muscle tissue during exercise?

9. What **change** occurs in the percentage of blood supply to the **abdominal organs** during exercise?

    Considering the **change in the percentage of blood flow** to the abdominal organs shown in the graph, what probably happens to the **diameter of the blood vessels** that supply blood to the abdominal organs during exercise?

10. What **change** occurs in the **percentage of blood supply to the skin**?

    How is this **change** in blood flow **helpful to the body during exercise**?

Introduction to Forensic Biology

Objectives

After completing this exercise, you should be able to:

- identify and interpret basic fingerprint patterns
- use your knowledge of fingerprint patterns and details to analyze forensic evidence
- demonstrate an understanding of the A-B-O and Rh blood typing systems
- use antigen–antibody interactions to identify and type unknown blood samples
- apply your knowledge of basic forensic techniques to real-life situations.

CONTENT FOCUS

A scientist is similar to a detective. Both use scientific thinking to solve problems without clear solutions. Solving this type of open-ended problem requires you to develop your skills in forming hypotheses, gathering evidence by observation and experimentation, and drawing conclusions on the basis of your findings.

Forensic science is a good example of the practical application of the scientific method—in this case, science is used to analyze evidence and solve crimes. Forensic scientists use information from all branches of science (including chemistry, geology, physics, medicine, and, of course, **biology**).

This exercise provides an introduction to two of the basic techniques that are employed in forensic biology: **fingerprint analysis and blood typing**.

ACTIVITY 1   GETTING STARTED

Preparation

Fingerprinting is the most commonly used and reliable system of identification. Each person has a unique pattern of ridges and valleys on their fingertips. Although other physical characteristics (such as weight, height, and hair color) may change as a person ages, fingerprints remain the same for life. Fingerprints penetrate through five skin layers and so are almost impossible to erase. Criminals have tried filing, acid burning, and even surgical removal, but with limited success.

Fingerprints have been used for identification purposes for several thousand years. The Babylonians recorded their fingerprints in the soft clay of their writing tablets (paper was not yet in use). The fingerprints served as a "signature" to prevent document forgeries. A similar system was also used in ancient China and Japan. In the last 100 years, an internationally recognized system of identifying fingerprints has been developed. Most law enforcement agencies (even in local communities) maintain their own fingerprint files. They can also access fingerprint information from state agencies and the FBI.

The skin is covered with **sweat glands**, including some in the ridges on the fingers and palms. During normal activities, perspiration from the sweat glands accumulates in these ridges. These ridges also accumulate body oils (from touching oil-producing areas such as the face, scalp, or neck). When you touch an object, perspiration and skin oils are transferred to that surface, leaving an invisible impression (the **latent fingerprint**). Latent prints found on nonabsorbent surfaces (such as metal or glass) can be dusted with colored powder and removed with transparent tape. Latent prints on absorbent surfaces (such as cloth or paper) must be recovered chemically.

The **Fingerprint Classification System** has three basic patterns: **arches, loops, and whorls**.

1. Begin by learning to recognize the basic identifying patterns used in fingerprint analysis.

Pattern area	Part of the fingerprint that is used for identification (**Figure 13-1a**)
Ridgelines	Individual lines that make up a fingerprint
Delta	Area where **two ridgelines** come to a **point** (**Figure 13-1b**)
Core	Approximate center of the finger impression (**Figure 13-1b**)
Whorl	**Complete circle** formed by ridgelines between two deltas; whorls **do not** continue off either side of the print (**Figure 13-2**)

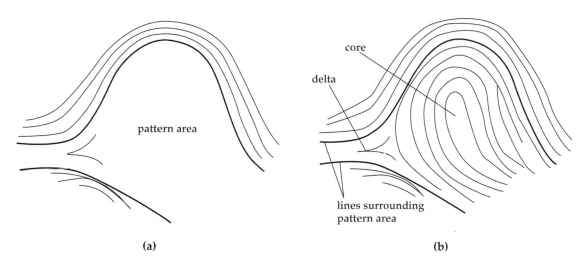

(a)   (b)

FIGURE 13-1.   (a) Fingerprint Pattern Area and (b) Features Used in Identifying Fingerprints

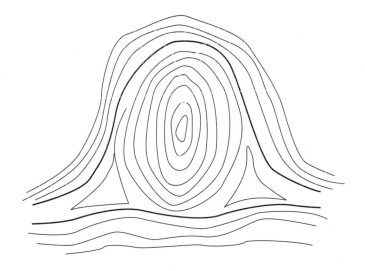

FIGURE 13-2.   Fingerprint with Whorl

✓ Comprehension Check

1. Refer to **Figure 13-1b**. How many **ridgelines** are between the delta and the core?

    _____

2. **(Circle one answer.)** The fingerprint in **Figure 13-1b is/is not** a whorl.

3. Refer to **Figure 13-2**. How many **ridgelines** are between the delta and the core?

    _____

2. In addition to whorls, there are several other common fingerprint patterns.

    Simple arch   Ridgelines **enter** on the **left**, rise, and **exit** on
                                     the **right** of the print (**Figure 13-3a**)

    Tented arch   Ridges meet at the center and form a **peak**
                                     (**Figure 13-3b**)

(a)

(b)

FIGURE 13-3.   (a) Simple Arch and (b) Tented Arch

3. There are various types of loops.

Loops	Curved ridgelines that **enter** and **exit** on the **same side** of the print **(Figure 13-4a)**
Double loop	Two separate loop formations on the same finger, between two deltas **(Figure 13-4b)**

(a)   (b)

FIGURE 13-4. (a) Loop and (b) Double Loop

4. Fingerprints that do not fit easily into any of the three categories are sometimes separated into a fourth category, called **mixed**. Mixed prints usually show a combination of two or more different patterns.

5. If additional identifying characteristics are required to make a fingerprint match, the following can be considered.

Bifurcation	Forking or dividing of one ridgeline into two or more branches **(Figure 13-5)**
Divergence	Spreading apart of two ridgelines that have been running parallel or nearly parallel **(Figure 13-5)**

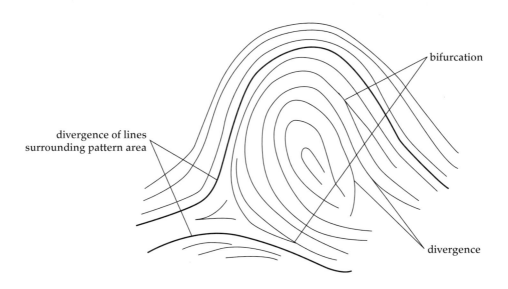

FIGURE 13-5. Additional Identifying Features

✔Comprehension Check

1. Place an **"X"** in front of each feature that is found in the fingerprints in **Figure 13-6**.

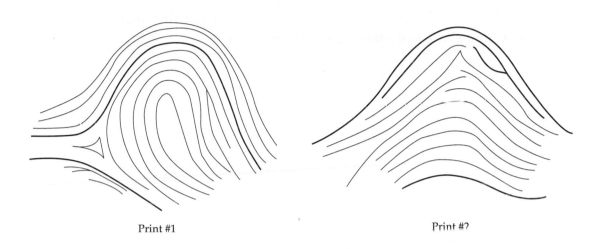

Print #1   Print #2

X	Characteristic
	Delta
	Whorl
	Simple arch
	Tented arch
	Ridged arch
	Loop
	Double loop
	Mixed
	Divergence of ridgelines
	Bifurcated ridgelines

X	Characteristic
	Delta
	Whorl
	Simple arch
	Tented arch
	Ridged arch
	Loop
	Double loop
	Mixed
	Divergence of ridgelines
	Bifurcated ridgelines

FIGURE 13-6. Analysis of Fingerprints

Check your answers with your instructor before you continue.

ACTIVITY 2   THE FINGERPRINT "FORMULA"

Preparation

A fingerprint formula is a list of **print classifications for each hand.** Each hand has a distinctive fingerprint formula. Before you can analyze crime-scene prints, you must develop some expertise in recognizing the different fingerprint patterns.

1. Work in groups. Get the following supplies: **one fingerprint inkpad per group, a magnifying glass, and some paper towels.**

2. **Remove Figure 13-7** from your notebook and place it on the table in front of you. Using the fingerprint inkpad, make a clean set of fingerprints for each member of your team.

    Place the prints on the **Fingerprint File Card** in **Figure 13-7**. To make the prints, place **one side** of each finger on the inkpad. **ROLL** the finger from one side to the other, covering the **FRONT** (**NOT** the tip) with ink.

Note:
Be careful with the fingerprint ink. It can stain your clothes.

---

3. **LIGHTLY** repeat the rolling motion to transfer the inked print onto **Figure 13-7** (or a clean sheet of white paper).

    If you press too hard, you will get a black ink blob with no discernable ridges. Use a very light touch to obtain a readable print.

4. To prepare an individual's fingerprint formula, **begin at the thumb and proceed to the little finger.** For example, a hand that has fingers of arch, arch, whorl, loop, whorl has a print formula of **a-a-w-l-w**.

    Using your own fingerprints, **determine your fingerprint formula. Write the fingerprint formula directly underneath each fingerprint in Figure 13-7.**

Left Hand   Thumb

Right Hand

Thumb

FIGURE 13-7. Fingerprint File Card

5. **By group discussion, double-check** each person's assessment of their own finger-print formula.

   Hard-to-classify prints may require your group to clarify and add information to the definitions of each category. As the distinctions between print patterns become more clearly defined, it will be easier for you to assign individual prints to a print type.

 **Check your answers with your instructor before you continue.**

6. Using your revised print type descriptions, **classify the twelve fingerprints in Figure 13-8**.

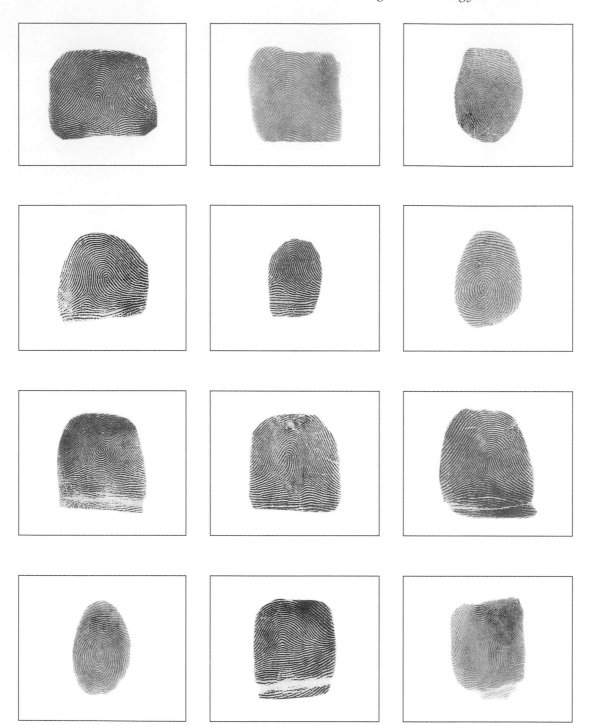

FIGURE 13-8. Practice Fingerprints

Check your answers with your instructor before you continue.

ACTIVITY 3   DETERMINING POINTS OF SIMILARITY

As you can see from looking at the fingerprints in **Figure 13-8,** there are only a few possible fingerprint patterns. Consequently, it is necessary to look at the **fine details of the ridgelines** within the prints (such as bifurcations, divergences, etc.) to match sample fingerprints to a specific person.

Exact matches in ridgeline patterns between two prints are called **points of similarity**. For a conclusive match between two prints, a minimum of six points of similarity is usually required. Since you are just learning how to do this, **one or two points of similarity per fingerprint** will be sufficient.

Circle each point of similarity you find, mark it with a letter designation, and provide an explanation for each point, as shown in **Figure 13-9**.

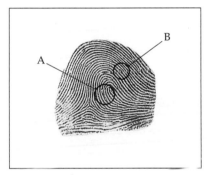
Sample print 1

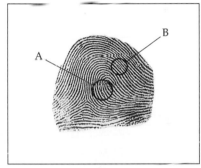
Sample print 2

Explanation
Both prints are right-facing loops

A   Bifurcation forming a broken oval
        pattern with a complete oval pat-
        tern below

B   Oval pattern in ridgeline

FIGURE 13-9. How to Label Points of Similarity

✔ Comprehension Check

Mr. Akumba arrives home and discovers that his back porch window has been forced open. He looks around the house and discovers that his new TV is missing. When the police arrive, they dust the scene for fingerprints. A few days later, Ms. Wright is arrested while breaking and entering another house in the neighborhood. You are the fingerprint examiner for the police department. The chief detective has asked you to compare Ms. Wright's fingerprints with those found at Mr. Akumba's house (**Figures 13-10** and **13-11**).

Left Hand   Thumb

Right Hand

Thumb

FIGURE 13-10.   Fingerprint File Card for Ms. Wright

FIGURE 13-11.   Prints Lifted from Mr. Akumba's Window

1. Do any of the prints from the crime scene match those of the suspect? If so, which prints match which fingers? List the matches below.

2. List the **points of similarity** you used to make your match. Mark each point with a letter on the fingerprint and provide an explanation for each letter (as shown in **Figure 13-9**).

3. **Did Ms. Wright do wrong?** Were those prints at Mr. Akumba's house hers? In a few sentences, summarize the conclusion you reached on the basis of your fingerprint examination.

Check your answers with your instructor before you continue.

ACTIVITY 4 — BLOOD TYPING

Scientists have identified and studied many different human blood groups. Blood groupings are based on the presence of **identifying proteins (also called antigens)** that are incorporated into the cell membranes of all body cells. Although red blood cells have **no personal identifying proteins**, they do have antigens that give us our different blood types. These molecules are referred to as antigens because they are identified as foreign if placed in the body of a person who does not have these molecules.

The most commonly known antigens are those that determine the **A-B-O and Rh** blood groups. These blood groups are well known because they are used to type blood for transfusions and organ transplants. The groups include the following:

- **Type A** blood contains antigen **A** and forms antibodies against **B**.

- **Type B** blood contains antigen **B** and forms antibodies against **A**.

- **Type AB** blood contains both antigens **A and B**. Type AB **does not form antibodies against either A or B.**

- **Type O** contains **no A or B antigens** but forms antibodies against **both A and B.**

- The **Rh factor** is the name given to another antigen present in red blood cells. People with **Rh+ blood have the antigen**, while those with **Rh– blood do not.**

Naturally occurring **antibodies (defensive proteins)** in the blood cause a serious transfusion reaction called **agglutination** if a person is transfused with blood containing **foreign** antigens of the A-B-O or Rh blood groups. Babies are not born with these specialized antibodies, but they begin to develop them a few months after birth.

The combination of an antigen and its corresponding antibody produces agglutination (clumping of blood cells). The agglutination (clumping) reaction makes it possible to determine blood type.

For example, type A blood has the A antigen. If we add **anti-A** (an antibody that reacts to antigen A) to this blood sample clumping will occur. The addition of **antibody-B**, however, would not cause these cells to clump. By analyzing the reactions of an unknown blood sample with specific antibodies, the blood type can be determined.

1. Work in groups. Get the following supplies: **a spot plate, a wax pencil, some toothpicks, a magnifying glass, and dropper bottles of the following: AB– blood, O+ blood, anti-A serum, anti-B serum, and anti-Rh serum.**

> ### Note:
> **These tests are being carried out with *simulated blood*, which is nonbiological, and nontoxic.**

TABLE 13-1 CLUMPING REACTIONS THAT OCCUR IN BLOOD TYPING		
BLOOD TYPE	EFFECT OF ADDING ANTIBODY-A	EFFECT OF ADDING ANTIBODY-B
A	Clumps form	No clumps
B	No clumps	Clumps form
AB	Clumps form	Clumps form
O	No clumps	No clumps

2. You will be testing each blood sample for the presence of the antigens A, B, and Rh.

    With the wax pencil, label **three horizontal rows** of your spot plate **A, B, and Rh** as shown in the sample in **Figure 13-12**.

3. You will be practising on two blood samples of **KNOWN type (AB and O+)**. With these two samples as controls, you will be able to see exactly what the clumping patterns look like for each possibility of A, B, and Rh antigens.

    Label the **first two vertical rows** of your spot plate **AB–** and **O+** as shown in **Figure 13-12**.

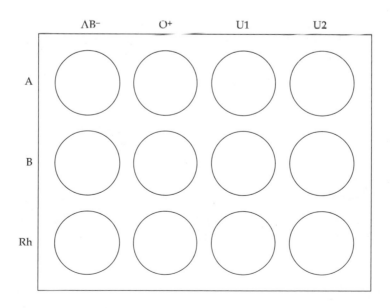

FIGURE 13-12. Sample Spot Plate

4. Place one large **drop of anti-A solution** into each of the **depressions in the row labeled A (one for the known sample of AB– blood and one for the known sample of O+ blood)**.

   Place one large **drop of anti-B solution** into each of the **depressions in the row labeled B**.

   Place one large **drop of anti-Rh solution** into each of the **depressions in the row labeled Rh**.

5. For **each of the three depressions for known sample AB–**, put a drop of blood from the **dropper bottle labeled AB–**.

   Stir each sample with a toothpick. **Use a CLEAN TOOTHPICK for each sample.**

6. For **each of the three depressions for known sample O+**, put a drop of blood from the **dropper bottle labeled O+**.

   Stir each sample with a toothpick. **Use a CLEAN TOOTHPICK for each sample.**

7. Observe the mixtures in the six depressions.

   If the agglutination test is positive, you will see a collection of small clumps, or particles, in the depression.

   You may find this easier to see by using a magnifying glass or dissecting microscope.

8. Return to the diagram in **Figure 13-12**. With a colored pencil or a highlighter, **color** each depression where you observed a **clumping** reaction.

9. Repeat the same test procedures using the **two unknown blood samples** supplied by your instructor.

   Label the **next two vertical rows** of your spot plate **U1 and U2** to represent the two unknown samples.

> ### Note:
> **Do not empty the liquid from the depressions containing the positive and the negative controls.**

10. The blood type of unknown **sample U1** is _____.

    The blood type of unknown **sample U2** is _____.

    Explain your answers.

✔ Comprehension Check

Owing to a clerical error, several samples of blood stored at the local blood bank may be incorrectly labeled as to the blood type. The three tests listed below were conducted on each sample. Use your knowledge of antigen–antibody reactions to sort the samples into their correct blood types

Test 1   Unknown sample mixed with "anti-A" serum (contains "type A" antibodies)

Test 2   Unknown sample mixed with "anti-B" serum (contains "type B" antibodies)

Test 3   Unknown sample mixed with "anti-Rh" serum (contains antibodies against the Rh protein)

1. **Unknown Sample A:** When tested, agglutination (clumping) of red blood cells occurred in Tests 1, 2, and 3.

   Blood Type: _____

2. **Unknown Sample B:** When tested, no agglutination occurred in any of the three tests.

   Blood Type: _____

3. **Unknown Sample C:** When tested, agglutination occurred in Tests 2 and 3, but not with Test 1.

   Blood Type: _____

4. **In your OWN words**, explain the chemical reaction that occurs when anti-B serum is mixed with

   a.   a type A blood sample:

   b.   a type B blood sample:

   c.   an Rh-positive blood sample:

Note:

Blood typing can be useful; however, it is not highly informative. Since there are only four blood groups in the A-B-O system, many people share the same type. Even if the blood type of the evidence matches the suspect, it does not prove that this blood came from the suspect. It could be from anyone that has the same blood type. For this reason, forensic scientists have moved away from conventional blood typing toward the more specific DNA typing technology.

Self Test

Fill in the blanks with the choice that is **most appropriate.** Answers can be used **only once.**

a. Ridgeline   f. Bifurcation
b. Delta   g. Divergence
c. Whorl   h. Fingerprint formula
d. Tented arch   i. Points of similarity
e. Loop   j. Sweat glands

1. ___ Complete circle formed by ridgelines between two deltas.

2. ___ Two parallel ridgelines spread apart.

3. ___ Ridgeline divides into two or more branches.

4. ___ Curved ridgelines that enter and exit on the same side of the print.

5. ___ Produce fluids that form latent fingerprints.

6. ___ Set of print classifications for each hand.

7. ___ Exact matches of ridgeline patterns between two fingerprints.

8. ___ Fingerprint pattern in which ridges meet at the center to form a peak.

9. Why do your fingers leave prints when you touch something?

10. Why is it unlikely that identical twins would have identical fingerprints?

Identify the most likely blood type on the basis of the results of the following antigen–antibody reactions. **Explain each answer.**

11.   When tested, agglutination occurred when exposed to anti-A serum, but not with anti-B or anti-Rh serums.

12.   When tested, agglutination occurred when exposed to anti-A serum, and also with anti-B serum. No agglutination occurred with anti-Rh serum.

13.   When tested, agglutination did not occur when exposed to both anti-A and anti-B serums. Agglutination did occur with anti-Rh serum.

EXERCISE

14

Mitosis and Asexual Reproduction

Objectives

After completing this exercise, you should be able to:

- name and describe the stages of mitosis
- identify the stages of plant and animal mitosis as viewed through the compound microscope
- discuss the relationship between the number of cells in each of the stages of mitosis and the length of the various stages
- explain the relationship between mitosis and the processes of regeneration and asexual reproduction
- apply your knowledge of asexual reproduction to cloning and other medical technologies.

CONTENT FOCUS

Most of us are aware that the outer layer of our skin is subject to constant wear. To keep this protective layer intact, skin cells must be replaced throughout our lives.

Cells of the epidermis, for example, are replaced every 25–45 days. A similar situation occurs elsewhere in the body.

Growth is another situation in which additional body cells are required. A **zygote** (fertilized egg) begins life as a single cell that multiplies into many cells as the embryo develops. During childhood, cell multiplication provides the many cells needed for growth to an adult body size.

The type of **cell division** that makes growth and repair possible is called **mitosis**. All body cells produced by mitosis must contain **the same genetic information** as all other body cells—the information that makes you a unique individual.

251

ACTIVITY 1   HOW MITOSIS WORKS

Preparation

A cell is genetically "programmed" to carry out its function in the body. These cellular instructions are included within structures referred to as **chromosomes**. A full set of these genetic instructions is necessary if a cell is to function normally. Therefore, when new body cells are produced by **mitosis**, each has a complete set of chromosomes.

The normal number of chromosomes in a cell is referred to as the **diploid number (abbreviated 2n)**. Each type of organism has a characteristic number of chromosomes. The diploid number of chromosomes in human body cells, for example, is 46. The diploid number in one species of pine tree is 24, in crayfish it is 200, and in fruit flies only 4.

Assume that the diploid number of chromosomes in the cell below equals 16.

If this cell divides, how many chromosomes must be present in each new cell?

Enter this number in each of the blank circles below.

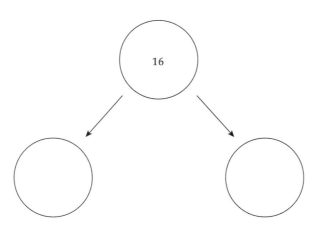

To make it easier to understand, the process of cell division is studied in five stages (called the **cell cycle**). At the completion of the cell cycle, two new cells are formed. Each cell will have a complete set of genetic instructions.

To complete cell division successfully, **dividing cells must solve several problems:**

Each new cell needs a full set of **chromosomes**.

Accomplished by: Duplicating chromosomes (an exact copy is produced).

To avoid confusion, these duplicate chromosomes are referred to as **sister chromatids**.

A cell preparing to divide will contain two complete sets of sister chromatids (one set for delivery to each of the two new cells).

Sister chromatids must be attached together.

Accomplished by: A structure called a **centromere** fastens the duplicates together.

Attachment makes it easier to keep track of sister chromatids for sorting into two groups (one set for each new cell).

Pieces of chromosomes should not get broken off or lost.

Accomplished by: The long, threadlike chromosomes coil and fold into compact structures. Condensed chromosomes are thick enough to be visible with the compound microscope.

The pairs of sister chromatids must be pulled apart and delivered to each new cell.

Accomplished by: A structure called the **mitotic spindle** hooks on to each pair of sister chromatids. **Spindle fibers** pull the chromatids apart and deliver one chromatid from each pair to each new cell.

When the sorting and delivery process is complete, the two cells must be separated into two **daughter cells**, each with its own set of chromosomes.

Accomplished by: A process called **cytokinesis** separates the cytoplasm into two halves (**"cyto"** means *cell* and **"kinesis"** means *cutting*).

During cytokinesis in animal cells, the cell membrane pinches into a groove called the **cleavage furrow**.

ACTIVITY 2   RECOGNIZING THE STAGES OF MITOSIS

Preparation

Use the **Decision Tree in Figure 14-1** to **identify the stages of the cell cycle** in **Figure 14-2.**

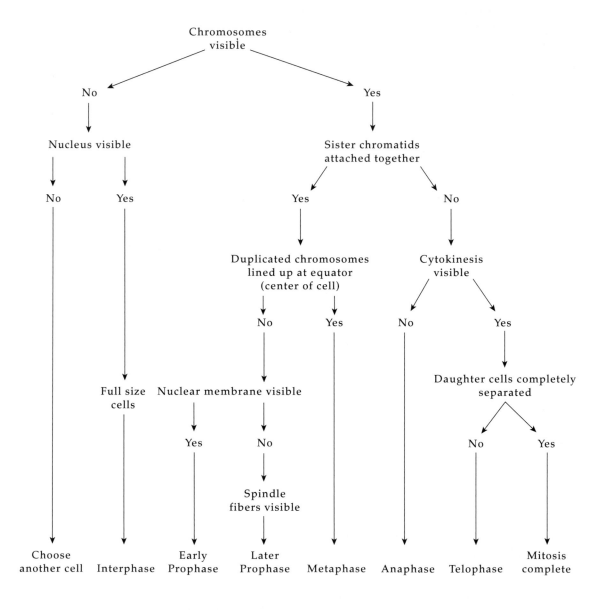

FIGURE 14-1.   Mitosis Decision Tree

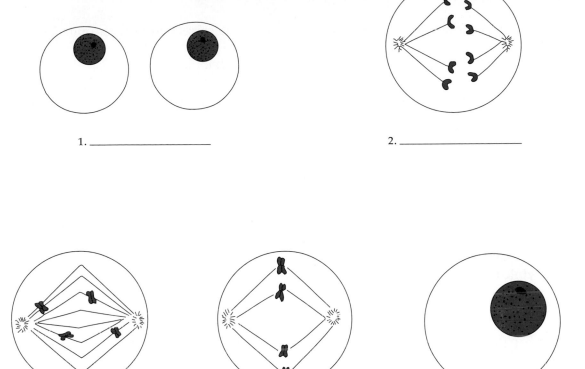

1. _____

2. _____

3. _____

4. _____

5. _____

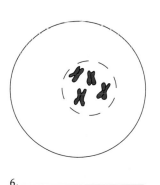

6. _____

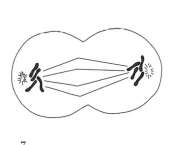

7. _____

FIGURE 14-2. Stages of the Cell Cycle to Identify

✓ Comprehension Check

1. Put the following stages of the cell cycle in the correct order (1 through 5).

    _____ Metaphase   _____ Telophase

    _____ Interphase   _____ Prophase

    _____ Anaphase

Use the following vocabulary words to fill in the blanks in the questions below. Each answer can be used **only once**.

centromere   **sister chromatids**   **cleavage furrow**
cytokinesis   **chromosomes**   **mitotic spindle**
equator

1. During metaphase, chromosomes line up along the _____ of the cell.

2. The two _____ are held together with a _____.

3. _____ refers to the division of the cytoplasm into two daughter cells.

4. Genetic information in cells is contained in structures called _____.

5. During anaphase, part of the _____ can be seen attached to the centromere of each sister chromatid.

6. The constricted appearance of the cell membrane prior to the formation of two daughter cells is called the _____.

Check your answers with your instructor before you continue.

ACTIVITY 3   MITOSIS—THE REAL THING!

Preparation

The stages of the cell cycle in living cells are often not as clear as in a set of diagrams. Sharpen your observation skills and practice recognizing the stages of mitosis at the same time.

1. For the exercise, **work on your own**. Get the following supplies: one **slide of mitosis in whitefish embryos** and a **compound microscope**.

2. View the slide with the **high-power lens (40×)**.

3. Your instructor will assign you **one stage of the cell cycle** to locate on your slide.

    When you locate a **clear** example of that stage, indicate the cell with the **pointer** in the ocular lens.

 Check your answers with your instructor before you continue.

4. Draw the stage **as it appears under the microscope**, in the appropriate location on **Figure 14-3**.

    In your drawing, label the following (if present): **chromosomes, mitotic spindle, equator, cleavage furrow, cell membrane, and nuclear membrane**.

5. After completing your drawing, **observe** the other **stages of the cell cycle** as identified by other students.

    Create your own drawings of these stages and add them to the appropriate locations on **Figure 14-3**.

    In your drawings, label the following (if present): **chromosomes, mitotic spindle, equator, cleavage furrow, cell membrane, and nuclear membrane**.

FIGURE 14-3.   The Cell Cycle as Seen Through the Microscope

ACTIVITY 4   ESTIMATE THE DURATION OF THE CELL CYCLE

Preparation

When you look at a prepared slide, you are looking at a "moment frozen in time." The preserving chemicals stopped the cells in the middle of their daily activities. From this slide, you can see how many cells were undergoing cell division at that moment. You can even see how many cells were in each of the stages of mitosis. You can use this information to estimate the length of each stage of the cycle compared to the other stages.

1. For this activity, **work individually**. Get the following supplies: one **slide of onion root mitosis** and a **compound microscope**.

   View the slide with the **high-power lens**. Position your view of the onion root tip to a section where many cells are undergoing mitosis (see **Figure 14-4** for examples).

   Your instructor will give you additional instructions on how to position the slide for optimal viewing.

2. Using the stages of onion root tip mitosis in **Figure 14-4** as a guide, **count** the number of cells **in your field of view** that are in each stage of the cell cycle.

   Record your results in **Table 14-1**.

TABLE 14-1 RESULTS OF CELL CYCLE OBSERVATIONS					
NUMBER OF CELLS IN STAGE	INTERPHASE	PROPHASE	METAPHASE	ANAPHASE	TELOPHASE
View 1					
View 2					
View 3					
Total for Views 1–3					
Percentage of Cell Cycle	%	%	%	%	%

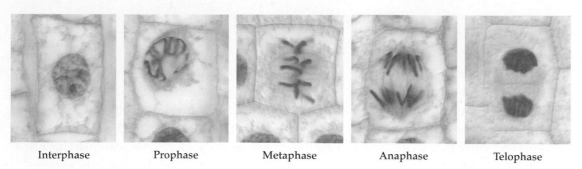

| Interphase | Prophase | Metaphase | Anaphase | Telophase |

FIGURE 14-4. Stages of Mitosis in the Onion Root Tip

3. Repeat the **instructions in Step 2 twice more** (so you will examine a total of **three different onion fields of view**) and record the results for each in **Table 14-1**.

4. **Total** the number of cells counted for each view and record the totals in **Table 14-1**.

5. **Calculate the percentage of cells observed in each stage of the cell cycle.**

STEP 1:

Add the total number of cells counted in all five stages.

Total Cells Counted = Interphase + Prophase + Metaphase + Anaphase + Telophase

Total Cells Counted = _____

STEP 2:

Divide the number of cells in Interphase by the total from Step 1 and multiply by 100.

$$\frac{\text{Number in Interphase}}{\text{Total Cells Counted}} \times 100 = \underline{\quad}\%$$

STEP 3:

Repeat Step 2 for the remaining four stages of the cell cycle. Record all percentages in **Table 14-1**.

6. Using the information in **Table 14-1**, form a **hypotheses** about the **length** of each stage of the cell cycle.

   Which do you think is the longest stage? _____

   Which do you think is the shortest stage? _____

   Do you think any stages are about the same length? If so, list them below.

   Explain how you formed your hypotheses.

7. See **Table 14-2 (after the Self Test)** for the results of some experiments on the length of the onion cell cycle. Do the experimental results support your hypotheses _____ **Explain your answer**.

ACTIVITY 5   REGENERATION

Preparation

Some animals have the ability to replace (regrow) lost or damaged body parts. This process is known as **regeneration** and is a form of asexual reproduction. You may be familiar with some animals that have this ability. Starfish (also known as **sea stars**) can regenerate body parts as long as a small part of the central disk is present. Starfish can grow a new arm and the severed arm can grow a whole new body! If cut in two, each half of the starfish can develop into a whole individual.

During the microscope exercise, you looked at planaria, freshwater flatworms. Planaria also have the ability to regenerate lost body parts. Any piece of the body except the tail can regenerate a complete new adult. You will cut **planaria** to demonstrate regeneration and regrowth.

1. Work in groups. Get the following supplies: **one clean razor blade in a plastic holder, a glass petri dish, a pipette, and crushed ice.**

2. You will also need a bottle of **pond water, labeling tape**, and a **dissecting micro-scope**.

    Set up your dissecting microscope, but keep the light **off**. On the stage of the micro-scope, place the **top half** of the petri dish.

    Fill the dish **to the rim** with crushed ice.

3. Place **one large planaria** into the **bottom half** of your petri dish with a small amount of **pond water**.

    Set the dish with the planaria **onto the crushed ice**. This will anesthetize the worm for the "operation."

    Let the worm relax on the ice for about **five minutes**.

4. While you are waiting, decide how you will cut the planaria. You can choose one of the two cutting patterns in **Figure 14-5**, or develop your own pattern.

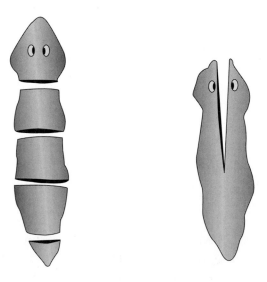

FIGURE 14-5. Planaria Cutting Patterns

5. Make **perpendicular cuts** with the razor blade through the worm. Do not tilt the razor blade when cutting. The cuts must be sharp and clean.

6. After the operation, add some clean **pond water** to the petri dish containing the worm.

   Discard the ice. Wash and dry the top half of the petri dish. Using labeling tape, mark the lid with **your laboratory section and your name**.

   Return the petri dish containing the worm to your instructor to be stored in a cool, quiet location.

7. After **two weeks**, you can examine your worm under the dissecting microscope to see the results of your experiment.

8. *Challenge Question!* If you cut a planaria into five pieces and each piece developed into a separate individual, would you be able to tell them apart? **Explain your answer.**

| ACTIVITY 6 | ASEXUAL REPRODUCTION: GROWING A PLANT FROM A CUTTING |

Preparation

As you have seen, mitosis provides genetically identical cells for growth, replacement, and repair of body tissues. Mitosis also provides a mechanism for **reproducing whole organisms asexually**. In asexual reproduction, offspring are produced, which are **genetically identical** to the single parent and also to one another.

Plants do not have to come from seeds. You can remove cuttings from your plants. The cuttings will develop roots and grow into new plants. This form of asexual reproduction is an example of **plant propagation**.

1. Get **a paper cup and a scalpel**.

   You will also find **plants for propagation, rooting hormone, and potting soil**.

2. Fill the cup with **potting soil** up to **one inch** from the top of the container.

    Moisten the soil with water until it is damp, but not soggy.

3. **Cut a small piece of stem** with several leaves attached.

    Dip the **cut end** of the stem into the **rooting hormone** and **plant** it in the cup.

4. Keep your baby plant **moist** and exposed to **bright, indirect light**. Roots should form within several weeks.

Comprehension Check

1. A botanist is trying to save a rare plant from extinction. Suggest a method she can use to increase the population.

2. The **type of cell division** that produced root growth in my cutting is _____.

3. Dolly, an internationally famous sheep, was conceived by an unusual process. Dolly was **cloned**—a type of **asexual reproduction**. What percent of Dolly's characteristics would be similar to her mother's? _____%. **Explain your answer**.

4. How would a colony of mice produced by **cloning be helpful** for testing the effectiveness of a new diet pill?

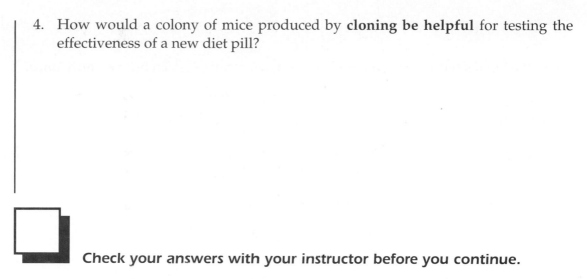

Check your answers with your instructor before you continue.

Self Test

Fill in the blanks with the most appropriate answer. Answers can be used **only once**.

a. Diploid   g. Cytokinesis
b. Sister chromatid   h. Cleavage furrow
c. Centromere   i. Equator
d. Spindle fibers   j. Zygote
e. Nuclear membrane   k. Daughter cell
f. Cell membrane   l. Chromosome

1. _____ Breaks down at the end of prophase, releasing the chromosomes.

2. _____ Chromosomes are arranged along this imaginary line during metaphase.

3. _____ Structure that holds two duplicate chromosomes together.

4. _____ Two of these are found at the **end** of mitosis.

5. _____ Normal number of chromosomes in a cell.

6. _____ Responsible for chromosome movement during mitosis.

7. _____ A constriction of the cell membrane that occurs in telophase.

8. _____ Duplicated chromosome has two of these.

9. _____ Process that separates the cytoplasm into two halves.

10. _____ Genetic material found in the cell nucleus.

Fill in the blanks with **true** or **false**.

11. _____ If a cell undergoing **mitosis** has a chromosome number of six, the daughter cells will have a chromosome number of three.

12. _____ In asexual reproduction, the offspring are identical to the parent.

13. _____ If cells are multiplying, as they do when grown in tissue cultures, mitosis is taking place.

14. _____ Growth during childhood is a good example of cell division.

15. Complete mitosis for the following parent cell. Draw each stage of mitosis **in order**, and show the number of chromosomes in each stage (including the two daughter cells). The beginning cell has a diploid number of four chromosomes.

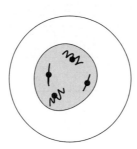

16. You are vacationing in Maine. One morning, while walking near the shore, you hear a group of fisherman complaining that there are too many starfish in the ocean. They plan to solve this problem by netting the starfish, cutting them in half to kill them, and throwing them back into the ocean. Is their plan likely to be effective? **Explain your answer.**

17. *Challenge Question!* If you wanted to asexually propagate a human, you could use

    a. a skin cell
    b. a sperm cell
    c. a red blood cell
    d. any of the above would be suitable
    e. none of the above would work out

Explain your answer.

TABLE 14-2

RESULTS OF EXPERIMENTS ON THE LENGTH OF STAGES IN THE ONION CELL CYCLE

	INTERPHASE	PROPHASE	METAPHASE	ANAPHASE	TELOPHASE
Length of Stage	17 hrs	88 min	4 min	3 min	6 min
Percentage of Cell Cycle	91%	7.9%	0.4%	0.3%	0.5%

15

Connecting Meiosis and Genetics

OBJECTIVES

After completing this exercise, you should be able to:

- name and describe the stages of meiosis
- correctly use and understand the terminology associated with cell division and genetics
- demonstrate an understanding of the changes in chromosome number that occur during meiosis and fertilization
- compare meiosis I with meiosis II in terms of the position of the chromosomes in each stage, changes in chromosome number, and number of daughter cells produced
- explain the process and importance of crossing over between homologous chromosomes
- draw and complete Punnett squares and use them to determine genetic probabilities in monohybrid crosses
- solve genetics problems involving other modes of inheritance, including incomplete dominance, codominance, and multiple allele traits.

CONTENT FOCUS

Your body cells contain **46 chromosomes** within the nucleus. As you probably know, **half** of this genetic information (23 chromosomes) is inherited from your mother and **half** (the other 23 chromosomes) from your father.

The genetic information carried by the **gametes** (the sperm and egg), when incorporated into a fertilized egg, will determine all the physical and physiological traits of the offspring.

There is a special type of cell division that changes the chromosome number from the normal **diploid** number (46) to the **haploid** number (23) found in sperm and eggs. This special type of cell division is called **meiosis**.

269

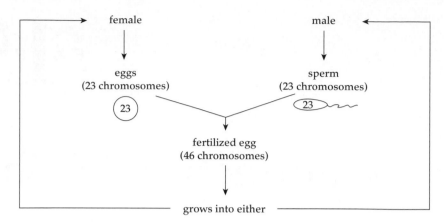

The stages of meiosis, in most respects, are similar to those of mitosis. The important differences include the following:

- the chromosome number is reduced from diploid to haploid to form gametes

- an exchange of genetic material takes place in a process known as **crossing over**; gametes produced by meiosis are **not** genetically alike

- meiosis involves two cell divisions (meiosis I and meiosis II); one occurs immediately after the other.

The genes on a chromosome may exist in **more than one form**, called **alleles**. Individuals inherit **two alleles for each trait**, one received from the mother in the egg and the other from the father in the sperm.

If the two inherited letters represent **different** forms of the gene (different **alleles**), an individual is called **heterozygous for that trait**. The allele that is **expressed in heterozygous individuals** is referred to as **dominant** and is represented by a capital letter (**B**).

The allele **not expressed** in heterozygous individuals is called **recessive**. Recessive alleles are represented with lowercase letters (**b**).

If individuals inherit two identical alleles for a trait (**BB or bb**), they are said to be **homozygous for that trait**.

The combination of alleles for a trait is an individual's **genotype** (such as BB or Bb).

The physical description of a specific trait is called the **phenotype** (such as brown eyes, type AB blood, or freckles).

In this exercise, you will work through the process of meiosis to **form** male and female **gametes, decode** the genetic information carried by these gametes, and use that information to **build** your baby's face.

GETTING STARTED:
ACTIVITY 1   BUILD A PAIR OF CHROMOSOMES

1. Work in groups. Get the following supplies: a **bag labeled "diploid human genome, male"** or **"diploid human genome, female."**

    In the supply area you will find containers of **beads** of several colors and shapes (**representing alleles**) and magnetic **connectors** (representing **centromeres**).

    Although humans have 23 pairs of chromosomes, we will simplify the process by using only **one chromosome pair.**

2. Fill your bag with **four beads from each jar**. Note that there are two jars of beads of each color. In one jar, the beads are striped. In the other jar they are solid.

    Solid-colored beads represent **dominant alleles** and **beads with a stripe** represent **recessive alleles for the same trait.**

    Get **two red centromeres and two yellow centromeres.**

3. **Remove Table 15-1 from one group member's book and lay it flat on the laboratory table. As you draw each bead (representing the alleles of your maternal and paternal chromosomes), place it into the appropriate box in Table 15-1.**

    Bead selection is random! Don't look in the bag while you're drawing the beads.

    a. The first bead you draw will become part of the chromosome you inherited from your **mother (maternal** chromosome).

    b. The next bead of the **same color** you draw will become part of the chromosome you inherited from your **father (paternal** chromosome).

    c. Additional beads of the **same color** you draw will be *THROWN BACK INTO THE BAG.*

    d. Draw beads until you have drawn **one of each color or shape for the maternal chromosome and one of each color or shape for the paternal chromosome.**

4. When you have drawn all the beads you need for both chromosomes, you are ready to hook your alleles together.

    Take a **yellow centromere** and hook it into position **between the green and red beads** of your **maternal chromosome**.

    Keeping the beads in the **correct order**, attach the remaining beads to the chromosome.

TABLE 15-1

TRAITS ON YOUR CHROMOSOMES

Bead Color or Shape	Maternal Alleles	Paternal Alleles	Trait	Alleles (Genotype)
purple			face shape	FF or Ff—round ff—triangular
orange			hair	HH—curly Hh—wavy hh—straight
yellow			eye size	EE—large Ee—medium ee—small
blue			eye distance	DD—close together Dd—medium spacing dd—far apart
green			eyebrow shape	BB or Bb—thick bb—thin
red			eyelash length	LL or Ll—long ll—short
white			nose size	NN—big Nn—medium nn—small
pink			lips	GG or Gg—thick gg—thin
black			ear lobes	RR or Rr—free rr—attached
white oval			cleft in chin	TT or Tt—present tt—absent
white twisted			freckles	QQ or Qq—freckles qq—no freckles

5. **Repeat step 5** to connect the alleles of the **paternal chromosome** using the **red centromere**.

   Together these **two chromosomes (maternal and paternal)** are referred to as **homologous chromosomes**. Homologous chromosomes carry alleles for the **same traits** (face shape, eye size, etc.), although the genetic information is **not identical**.

6. Fill in **Table 15–2** with the genotypes and phenotypes of the traits on the chromosomes you have just constructed.

   Solid beads = dominant alleles. Striped beads = recessive alleles.

TABLE 15-2 WHAT ARE YOUR TRAITS?		
TRAIT	YOUR GENOTYPE	YOUR PHENOTYPE
face shape		
hair		
eye size		
eye distance		
eyebrow shape		
eyelash length		
nose size		
lips		
ear lobes		
cleft in chin		
freckles		

HOMOLOGOUS CHROMOSOMES
ACTIVITY 2   SEPARATE IN MEIOSIS TO FORM GAMETES

1. Draw a series of circles that are the same as those shown in **Figure 15-2**.

    Make the circles big enough for your chromosomes to fit comfortably.

2. Place your pair of chromosomes in the first circle. Your cell is now in **interphase**.

    What is the diploid number of chromosomes in this cell? _____

    How many chromosomes came from your mother? _____ How many from your father? _____

    How many **traits** are represented on your chromosomes? _____

    How many **alleles** are represented on your chromosomes? _____

3. An important event that occurs during **interphase** involves the replication of chromosomes. Use the spare beads in your genome bag to **replicate** your two chromosomes.

    Two identical (replicated) DNA strands are called **sister chromatids**.

    Attach the new **sister chromatids** for your **maternal chromosome** together at the centromeres. Do the same for your replicated **paternal chromosome**.

4. Move your duplicated chromosomes to the next circle, marked **prophase** through **telophase of meiosis I**.

    You will complete **prophase through telophase of meiosis I** in the **same circle**.

5.   As shown in **Figure 15-1**, during **prophase I**, the **two homologous chromosomes** find each other and pair up in a process called **synapsis**.

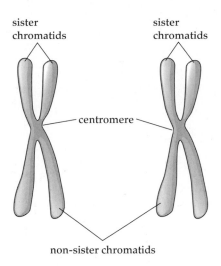

sister
chromatids

sister
chromatids

centromere

non-sister chromatids

FIGURE 15-1.   Duplicated Homologous Pair (Tetrad)

6.   During synapsis, the **non-sister chromatids** exchange genetic information. This process is called **crossing over**.

Simulate crossing over between **two non-sister chromatids** in your homologous pair (exchange of alleles between **one maternal and one paternal** chromosome).

Exchange alleles for the **last five traits** (nose size through freckles).

> ## *Note:*
> ### Exchange alleles ONLY for two NON-SISTER chromatids!

7.   Simulate **metaphase I** by placing the chromosomes **in the correct position** relative to the **equator**.

To simulate **anaphase I**, separate the two homologous chromosomes by moving them to the opposite poles of the cell.

To simulate the division of the cytoplasm in **telophase I**, draw a dotted line that represents the separated **daughter cells**.

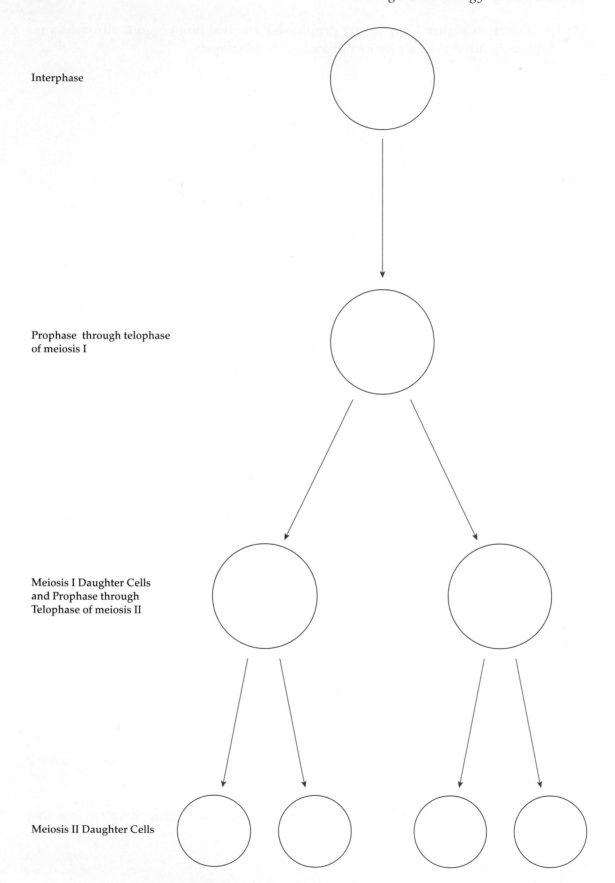

Interphase

Prophase through telophase
of meiosis I

Meiosis I Daughter Cells
and Prophase through
Telophase of meiosis II

Meiosis II Daughter Cells

FIGURE 15-2. Pattern for Meiosis Simulation

8. Move your chromosomes to the next two circles that are labeled **Meiosis I daughter cells.**

✓ *Comprehension Check*

1. What is the number of chromosomes in each daughter cell? _____

2. Does this number represent the **diploid** or the **haploid** chromosome number? _____

3. What happened to the **maternal and paternal** chromosomes from the original parent cell?

   Are the two daughter cells genetically identical? _____ **Explain your answer.**

9. Meiosis continues with a second cellular division.

   Simulate meiosis II without moving your chromosomes to another set of circles. Note that the circles you are now using have two labels. In **addition** to Meiosis I daughter cells, the circles are also labeled **prophase through telophase of meiosis II**.

   Nothing unusual occurs to the chromosomes in prophase II.

   To simulate **metaphase II**, place the chromosomes **in the correct position** relative to the **equator**.

   To simulate **anaphase II**, separate the two **sister chromatids** by moving them to the opposite poles of the cell.

   Move your chromosomes to the four circles that are labeled **Meiosis II daughter cells.** Each of these daughter cells has the potential to develop into a sperm or egg.

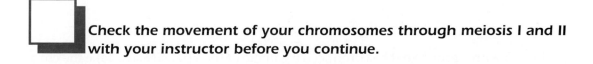

Check the movement of your chromosomes through meiosis I and II with your instructor before you continue.

✔ *Comprehension Check*

1. The original parent cell in your meiosis simulation had two alleles for each trait.

    How many alleles for each trait are **now** in each daughter cell? _____

2. Are your daughter cells diploid or haploid? _____

3. Are the four daughter cells **genetically identical**? _____ **Explain your answer**.

4. Where in the body does meiosis occur in **males**? _____ Where in **females**? _____

ACTIVITY 3   FERTILIZATION: NATURE'S
                      EQUIVALENT TO ROLLING THE DICE

1. As you know, **only one sperm and one egg** can participate in fertilization.

    Any of the four sperm can potentially fertilize an egg, but the development of female gametes is slightly different. **Only one** of the four daughter cells will develop into an egg cell. The other three are not functional and are referred to as **polar bodies**.

    You will simulate fertilization of your gametes with a roll of the dice. Get a **single six-sided die**.

    Are **your** gametes potential sperm or eggs? _____

2. Choose the gamete that will participate in fertilization as follows:

    a. **Number your daughter cells** 1 through 4.

    b. **Roll the die**. If you get a number between 1 and 4, that is your lucky gamete. If you get 5 or 6, roll again.

3. Link up with a group that has produced a gamete of the opposite sex. Take the chromosomes from both gametes and **place them in a circle** labeled "**Fertilized Egg**."

ACTIVITY 4   WHAT IS YOUR BABY'S GENOTYPE?

1. On the two chromosomes in **Figure 15-3**, list your baby's **alleles in their correct order**.

    Solid beads = dominant alleles. Striped beads = recessive alleles.

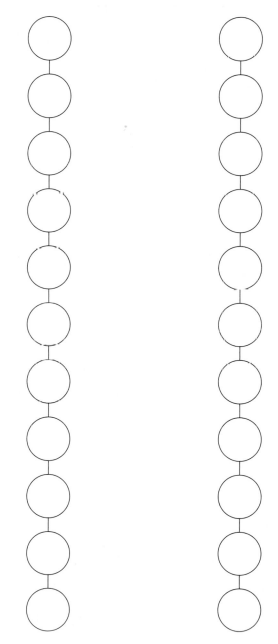

Maternal Chromosome   Paternal Chromosome

FIGURE 15-3. Maternal and Paternal Chromosomes of the Fertilzed Egg

ACTIVITY 5   GENOTYPE DETERMINES PHENOTYPE

1. Now that you know your baby's **genotype** for each trait, turn back to **Table 15-1**. In **Table 15-1, highlight or circle** your baby's **genotype and phenotype** for each trait.

    Now that you know your baby's **phenotype** for each trait, you are ready to see what your baby will look like!

2. Choose the appropriate **face shape** outline to begin your baby (**Figures 15-4** and **15-5**).

    Cut out the appropriate **eyes, ears, and other facial features** and use **tape** to attach them to your baby's face **(Figures 15-6 through 15-13)**. If your baby has freckles, **draw them** in.

    Add the appropriate **eyelashes** to your baby's face.

    Are you pleased with your results? If not, remember that each time a sperm fertilizes an egg you get a different combination of alleles. You may have better luck next time.

3. **Roll the die again** to determine if your baby is a boy or a girl. If you roll **1 through 3**, it is a **girl**. If you roll **4 through 6**, it is a **boy**.

FIGURE 15-4. Round Face

FIGURE 15-5.   Triangular Face

FIGURE 15-6. Curly Hair

FIGURE 15-7. Wavy Hair

FIGURE 15-8. Straight Hair

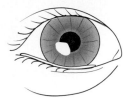

Large Eyes

Medium Eyes

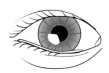

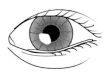

Small Eyes

FIGURE 15-9. Eye Types

Thick Eyebrows   Thin Eyebrows

FIGURE 15-10. Eyebrow Shape

Thick Lips   Thin Lips

FIGURE 15-11. Lip Types

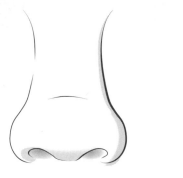
Large nose

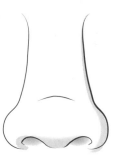
Medium nose

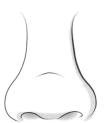
Small nose

FIGURE 15-12. Nose Size

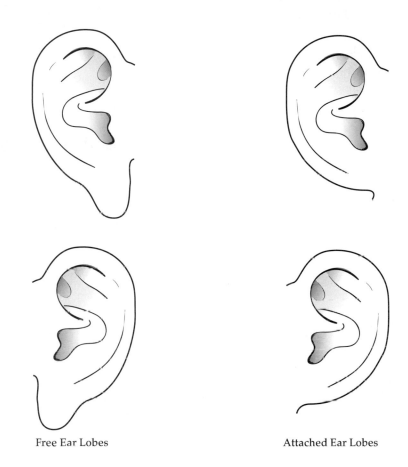

Free Ear Lobes   Attached Ear Lobes

FIGURE 15-13. Ear Lobes

ACTIVITY 6 PASSING ON TRAITS

Preparation

Twenty-four years have passed. Your baby is now an adult and has married someone who is **heterozygous for freckles**. What is the probability that they will have a child with freckles?

The first step in calculating the probability of freckles is to **determine the alleles carried by the sperm and eggs** of this couple.

As you know, alleles separate in meiosis, so that each sperm and each egg has only **one allele from each homologous pair**.

Your child's genotype for the freckle trait _____ _____

1. A **Punnett Square** is a convenient way to visualize the different combinations of alleles that might occur during fertilization.

2. The spouse's alleles have been entered on the top row of the Punnett Square. Enter the alleles for **your child** along the left side of the square.

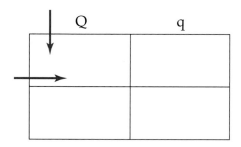

3. Using the Punnett Square as a guide, fill in the boxes by carrying the alleles down from the top and across (from left to right).

    Each box in the Punnett Square represents a **25% probability** of having a baby with that specific genotype.

    What is the probability that this couple will have a child with freckles? _____

4. **Assume that your baby has wavy hair** (inherited through **incomplete dominance**) and marries a person who **also has wavy hair**. Incomplete dominance is a special type of inheritance that occurs when an allele exerts only partial dominance over another allele. This results in an intermediate (blending) phenotype in heterozygous individuals. Complete the Punnett Square below to answer the following questions:

What is the probability of a child with **straight** hair? _____

What is the probability of a child with **curly** hair? _____

ACTIVITY 7   OTHER MODES OF INHERITANCE

Preparation

Now that you are beginning to be an expert in solving genetics problems, let us try something a little more challenging. Up to this point, we have dealt with traits that have **only two alleles**.

Human blood types (A, B, and O) are inherited by **multiple alleles**. There are **three possible alleles** for blood type, and **two** of those alleles are **codominant**.

Codominance is a special type of inheritance in which two alleles are equally dominant. Both alleles are expressed independently of each other, resulting in a heterozygous individual that shows both homozgous phenotypes. For example, red blood cells with both A and B proteins in their cell membranes.

Multiple alleles means that there are **more than two possible alleles** for that trait, although each person will inherit **only two** of those alleles (one from the mother and one from the father).

In human blood types, the alleles I^A and I^B are codominant. Both I^A and I^B are **dominant** over the **recessive allele i**.

GENOTYPE	PHENOTYPE
$I^A I^A$ or $I^A i$	type A blood
$I^B I^B$ or $I^B i$	type B blood
$I^A I^B$	type AB blood
ii	type O blood

1. When Howard Hughes (the reclusive multimillionaire) died, he left no legitimate heirs. Soon, however, a long succession of people claiming to be his children began to appear. A young man claiming to be Howard Hughes's child sued for a share of the estate. The judge ordered blood tests to determine the validity of the claim. Howard Hughes had blood **type AB**, the **mother** of the young man had blood **type A**, and the **young man** himself had **type O** blood. If you were the judge, how would you rule? **Explain your answer.**

2. Is it possible for a **type A** person married to a **type B** person to have **type O** children? **Explain** your answer.

Check your answers with your instructor before you continue.

Self Test

1. Complete the table by filling in the **missing stage** of the cell cycle or the **most important** cellular activity that would occur during that stage.

STAGE OF THE CELL CYCLE	CELLULAR ACTIVITY
Prophase I	
	Two haploid daughter cells first appear
Metaphase I	
Metaphase II	
	Chromosome replication
Anaphase II	
	Division of cytoplasm occurs
Telophase II	
	Homologous chromosomes separate
	Sister chromatids separate

2. **Albinism** is a **recessive** condition (aa) in which body cells cannot manufacture the pigment **melanin**, which colors eyes, hair, and skin. **Normal pigment production is dominant (AA or Aa)**. A normally pigmented woman marries a normally pigmented man. To their surprise, they have an albino child. Give the genotypes of the parents and the child:

    the father _____ the mother _____ the child _____

    What is the probability that this couple will have another albino child? _____

    What is the probability that they will have a normally pigmented child? _____
    Show your work.

3. **Tay-Sachs disease** in humans is controlled by a **recessive allele (t)**. This disease is characterized by the inability to produce an enzyme needed to metabolize lipids in brain cells. Without this enzyme, lipids accumulate in the brain cells and gradually destroy their ability to function. Affected children usually die by age five.

    What genotypes **must** be found in **both parents** in order to have a child with Tay-Sachs? _____ _____

    Explain your answer and **include a Punnett Square** that shows the possible genotypes among the children of this marriage.

4. Imagine you are a zookeeper in the Big Cat house of the National Zoo in Washington, D.C. Among the tigers, two yellow-coated parents (Ghandi and Sabrina) give birth to a cub named Snowflake that has a rare color variation—a white coat with black stripes.

    Assuming that the **white coat is inherited as a recessive allele (c)**, what are the genotypes of all the tigers?

    Ghandi _____   Sabrina _____   Snowflake _____

    If these parents have another cub, what is the probability that the cub will be **heterozygous** for the coat-color trait? **Show your work.**

5. Another zoo wants to start its own tiger exhibit. So far, they have one tiger in their collection, a white one. If the National Zoo sends Ghandi on breeding loan to this zoo, what is the probability that the zoo will get another white tiger for their new exhibit? **Show your work.**

EXERCISE

16

Useful Applications of Genetics

Objectives

After completing this exercise, you should be able to:

- demonstrate an understanding of the limitations of sample size in scientific data analysis
- determine genotypes using pedigree charts
- explain why more males than females express X-linked traits
- solve genetics problems involving dominant–recessive inheritance, X-linked traits, and codominance
- explain the evolutionary relationship between sickle-cell disease and malaria
- apply your knowledge of genetics to real-life situations.

CONTENT FOCUS

Not all alleles produce visible traits like skin color or height. Most alleles control **physiological** traits, such as production of digestive enzymes, hormones, and antibodies. Alleles are responsible for invisible traits such as blood type, ability to carry out metabolic pathways (such as producing proteins or storing blood sugar), color vision, and many others. One example of a physiological trait controlled by a **single gene** is the ability to taste a harmless chemical, **PTC** (phenylthiocarbamide).

ACTIVITY 1 — PTC TASTING

1. Get the following supplies: one piece of **control** taste paper and one **containing PTC.**

2. **Taste the control paper** first (to establish the taste of the paper itself). Then **taste the PTC paper.**

If you are a taster, you will detect a very unpleasant, bitter taste. If there is little difference between the PTC and the control papers, you are not a taster.

Are you a taster or a nontaster? _____

3. Record your taste-test results **on the master chart at the front of the room.**

4. In the general population, approximately 75% of people can taste PTC. The remaining 25% are not able to taste this chemical.

   Were your class results close to the **expected percentage?** _____

   If not, suggest a possible reason why your class results differed from the expected outcome:

5. The ability to taste PTC comes from a **dominant allele (T).** Using this information, fill in the appropriate genotypes for tasters and nontasters.

   Tasters _____   _____

   Nontasters _____

6. The inheritance pattern of traits like PTC tasting can be diagrammed in a chart called a **pedigree.** A pedigree illustrates the marriages for several generations within a family and the children produced.

   ■ **Females** are shown with **circles** and **males are shown** with **squares.**
   ■ A **black square** or a circle shows the **presence of the condition** being studied. A **white square or a circle** means the **condition is absent** in that person.
   ■ A marriage or mating is shown by a line connecting the parents.
   ■ Children from a mating are shown by a vertical line between the parents.
   ■ All individuals from the **same generation** are shown along the same horizontal line.

7. After examining the **pedigree key**, determine the genotypes of **all the people** in this family. In a few cases, there may not be enough information to determine a person's second allele. In this situation, **enter a question mark (?) in place of the second letter.**

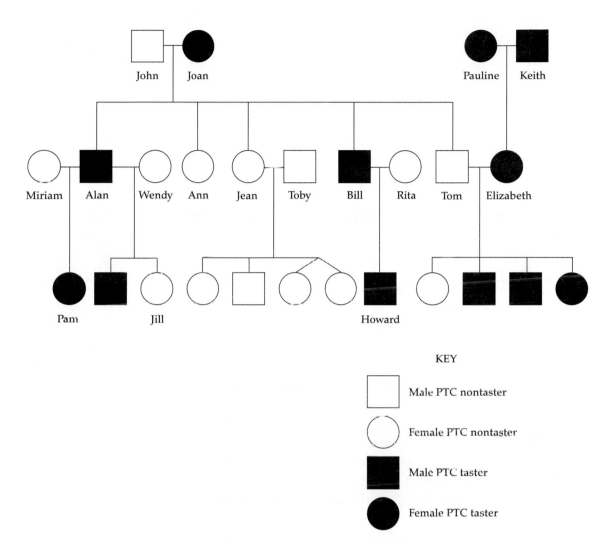

KEY

☐ Male PTC nontaster

○ Female PTC nontaster

■ Male PTC taster

● Female PTC taster

FIGURE 16-1. PTC Pedigree

8. Referring to the pedigree in **Figure 16-1**, if Jill marries Howard, what is the **probability** that their children will be

tasters? _____   nontasters? _____

Show your work!

9. If Pam marries Howard, what is the probability that their children will be: tasters? _____ nontasters? _____

Show your work!

Check your answers with your instructor before you continue.

ACTIVITY 2   TASTING OF OTHER CHEMICALS

Preparation

Since the ability to taste PTC is genetically based, it is likely that the ability to taste other substances can also be genetically determined. Our preferences for certain foods and dislike for others is determined, in part, by the types of chemicals we taste.

An example is the **artificial sweetener, Nutrasweet**, used in diet foods and beverages. Nutrasweet may taste **sweet or bitter** (depending on the individual). Another example is **sodium benzoate**, a **food preservative**. You may see it listed among the ingredients on the labels of many packaged foods. When sodium benzoate is tasted alone, tasters may describe it as **salty, sweet, sour, or bitter**.

Do you think there might be a relationship between tasting PTC and tasting sodium benzoate? Write a hypothesis about the relationship between PTC tasting and tasting sodium benzoate.

CHEMICAL TASTING HYPOTHESIS

Check your hypothesis with your instructor before you continue.

ACTIVITY 3   TESTING YOUR HYPOTHESIS

> ### Note:
> There are not enough people in your laboratory group to give you an adequate test sample. People from other groups will be happy to volunteer as test subjects.

1. Use **Table 16-1** to record your data.

TABLE 16-1 CHEMICAL TASTING DATA		
SUBJECT		
1		
2		
3		
4		
5		
6		
7		
8		
9		
10		
11		
12		

2. In a few sentences, **summarize the results** of your experiment.

3. Was your hypothesis **supported**? _____

4. **Write a conclusion based on your hypothesis and the collected data. Support
    your conclusion** by mentioning facts collected during your experiment.

Comprehension Check

1. After performing this experiment, do you think that the ability to taste certain
    chemicals can influence your food preferences? **Explain your answer.**

2. **Barium sulfate** is tasteless to some people but bitter to others. Two sisters are
    having their gastrointestinal tracts x-rayed. They are both asked to drink a
    barium sulfate "milk shake" before the procedure. One sister drinks the "milk
    shake" without complaint. The other sister complains that it tastes terrible.
    Using the information gained in today's lab activities, **suggest an explanation
    for this difference.**

ACTIVITY 4   X-LINKED TRAITS

Preparation

In humans, **the X chromosome is large in comparison to the Y chromosome**. The X chromosome carries information for many traits that are not related to the sex of the individual. Alleles carried only by the X chromosome are said to be **X-linked** (or sometimes, sex-linked).

Some of the alleles on the tiny **Y chromosome** appear to have no counterparts on X. These **Y-linked alleles** code for traits that are found **only in males**.

Among the X-linked traits are a number of recessive genetic disorders. One of these is **hemophilia**, the inability to produce proteins necessary for blood clotting. Hemophiliacs may bleed to death from relatively minor cuts or bruises. Historical records dating back thousands of years mention the inheritance pattern of hemophilia. Among the ancient Hebrews, sons born to women with a family history of hemophilia were excused from circumcision.

Hemophilia was common during the 1800s in the royal families of Europe, whose members often intermarried. Queen Victoria of England was a **carrier** of the trait. She had one X chromosome with the allele for normal blood clotting (X^H) and the other with the defective allele (X^h). Because she did have **one normal dominant allele**, her blood clotted normally.

Her husband, Prince Albert, was completely normal for this trait. He had one normal allele on the X chromosome (X^H) and a Y chromosome with **no allele related to blood clotting** (Y^o).

Eighteen of Queen Victoria's 69 descendants were carrier females or hemophiliac males. Crown Prince Alexis of Russia was one of these hemophiliac descendants. His affliction indirectly contributed to the overthrow of the monarchy in Russia.

1. Complete this Punnett Square for the marriage of Victoria and Albert.

Queen Victoria

	X^H	X^h
Prince Albert X^H		
Y^o		

2. What is the probability that Victoria and Albert could have

   a hemophiliac son _____
   a hemophiliac daughter _____
   a normal son _____
   a carrier daughter _____
   a daughter with normal blood clotting _____

3. Shown below is a partial pedigree of hemophilia in the descendants of Victoria and Albert. **Affected individuals have darkened circles or squares.** For **each individual in the pedigree,** fill in their probable genotype.

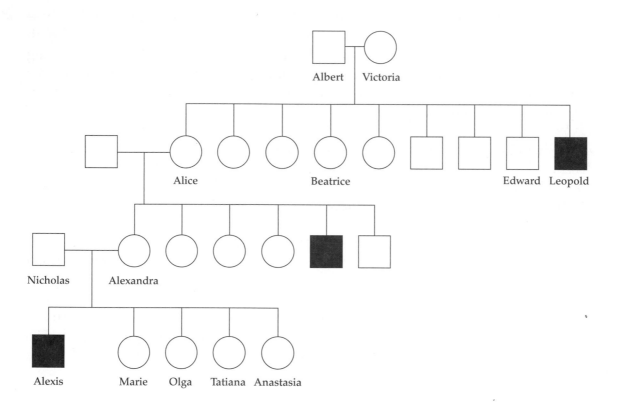

FIGURE 16-2. Partial Pedigree of Hemophilia in the Descendants of Victoria and Albert

✓ Comprehension Check

1. *Challenge Question!* If Victoria and Albert had a hemophiliac daughter, why would it be **unlikely for her to survive puberty** without medical intervention?

ACTIVITY 5   CODOMINANCE AND SICKLE-CELL ANEMIA

Preparation

Codominance is a special type of inheritance in which two alleles are equally dominant, and, therefore, heterozygous individuals show **both homozygous traits**.

Sickle-cell anemia is an example of a human trait that is inherited through codominance. This condition is quite common in countries that have a high incidence of malaria. Sickle-cell anemia is a blood disorder affecting oxygen transport by hemoglobin. The sickled cells block capillaries, depriving the tissues of needed oxygen.

Individuals with sickle-cell anemia frequently die at an early age. **Heterozygous** individuals have **sickle-cell trait.** These carriers are usually healthy, but experience some problems with intense exercise or under low-oxygen conditions. In the United States, sickle-cell trait affects 1 out of 12 African Americans.

The allele for normal hemoglobin is Hb^A and the sickle-cell allele is Hb^S. The following is a summary of the possible sickle-cell genotypes and phenotypes.

$Hb^A Hb^A$	completely normal
$Hb^A Hb^S$	sickle-cell trait (this person has a combination of normal hemoglobin and the abnormal, sickled form of hemoglobin)
$Hb^S Hb^S$	sickle-cell anemia (all abnormal hemoglobin)

1. What are the possibilities for children if **both parents are heterozygous** for sickle-cell trait?

2. What is the probability of a couple having a child with sickle-cell **trait** if **one** parent is **normal** and the **other** has **sickle-cell trait?**

ACTIVITY 6   A FAMILY HISTORY OF SICKLE-CELL DISEASE

Preparation

The foundation of our modern knowledge of sickle-cell anemia is based on the research of Dr. Angela Ferguson and Dr. Roland Scott, professors at Howard University in Washington, D.C. The doctors became interested in sickle-cell anemia because many of their friends and some family members were affected by the disease. They published the first research paper on sickle-cell disease in the 1940s—**25 years ahead** of other researchers.

Imagine that you are a physician studying blood genetics in the laboratory of Drs. Ferguson and Scott. The Minister of Health from Zaire has requested their help to investigate two health problems within the villages of his country that seem to be related: sickle-cell disease and malaria. Drs. Ferguson and Scott send you to investigate.

After spending a year looking into the problem, you have collected data about **Family A** and constructed a **pedigree and a health profile** for this family. **Each individual on the pedigree is identified by a number.**

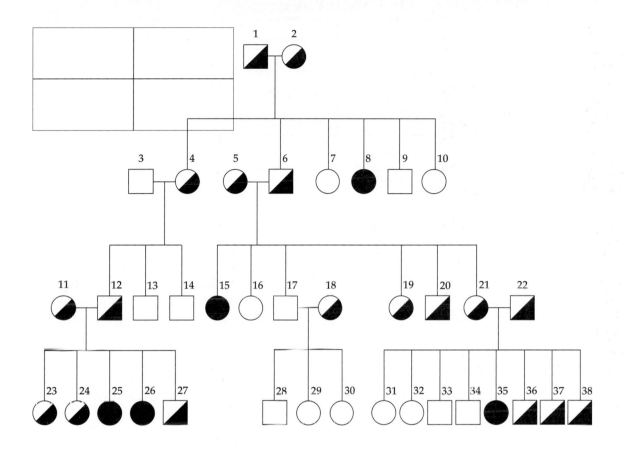

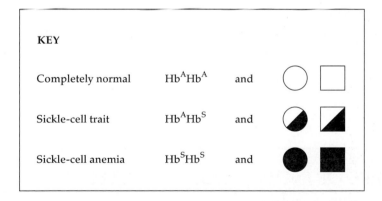

KEY

Completely normal	$Hb^A Hb^A$	and	◯ ▢
Sickle-cell trait	$Hb^A Hb^S$	and	◐ ◨
Sickle-cell anemia	$Hb^S Hb^S$	and	● ■

FIGURE 16-3. Pedigree of Family A

TABLE 16-2
HEALTH PROFILE OF FAMILY A

HEALTH STATUS	AFFECTED FAMILY MEMBERS
Affected with malaria	13, 32, 33
Dead from malaria	7, 9, 10, 16, 29, 30, 31, 34
Dead from sickle-cell anemia	8, 15, 25, 26, 35
Dead from causes unrelated to malaria or sickle-cell anemia	3, 19
In good health	1, 2, 4, 5, 6, 11, 12, 14, 17, 18, 20, 21, 22, 23, 24, 27, 28, 36, 37, 38

Malaria is one of the world's most serious diseases. It is transmitted by the bite of an infected mosquito. The bite allows a parasitic protozoan to be passed from the mosquito to the bloodstream of humans. Once in the body, the parasites live and multiply inside red blood cells.

Every two to three days, infected red blood cells burst, releasing thousands of new parasites that will infect even more red blood cells. The infected person experiences bouts of chills and fever each time such a release occurs.

1. Refer to the Health Profile of Family A in Figure 16-2. On the **pedigree chart in Figure 16-3, place an asterisk** (*) next to the numbers of all persons who are affected with malaria or who have died from malaria. **List** the pedigree number for **each** individual in Family A who has **died from malaria or is currently affected** with this disease.

TABLE 16-3
GENOTYPES OF INDIVIDUALS AFFECTED BY MALARIA

LIST PEDIGREE NUMBER FOR EACH AFFECTED INDIVIDUAL (SEPARATED BY GENOTYPE)	GENOTYPE
	Hb^AHb^A
	Hb^AHb^S
	Hb^SHb^S

2. **Of the total number of deaths in this family**, what **percentage** died from each cause? Write your answers in Table 16-4.

TABLE 16-4 **CAUSES OF DEATH IN FAMILY A**		
CAUSE OF DEATH	NUMBER DEAD	PERCENTAGE OF TOTAL DEATHS
Malaria		
Sickle-cell anemia		
Other		
TOTAL		

3. What is the **leading cause of death** among the members of Family A?

4. **Go back to the pedigree** and examine the genotypes of the individuals who are alive and in good health.

   Among these healthy individuals, what is the most frequent **genotype** for sickle-cell disease? _____

5. **Write a hypothesis** about the relationship between malaria and **one of the three genotypes** related to sickle-cell disease.

6. As you know, a hypothesis is an **educated** guess about a relationship. **Explain** how you **developed** your hypothesis by **listing facts** from the pedigree of Family A.

7. Is the death pattern shown by Family A typical for the population as a whole? To answer this question, **use Figure 16-4 to make a graph** comparing:

    a. the percentage of deaths from each cause in Family A (as shown in **Table 16-4**)

    b. the percentage of deaths from each cause in several neighboring communities (as shown in **Table 16-5**)

TABLE 16-5		
CAUSES OF DEATH IN NEIGHBORING COMMUNITIES		
CAUSE OF DEATH	NUMBER DEAD	PERCENTAGE OF TOTAL DEATHS
Malaria	1080	
Sickle-cell anemia	600	
Other	320	
TOTAL	2000	

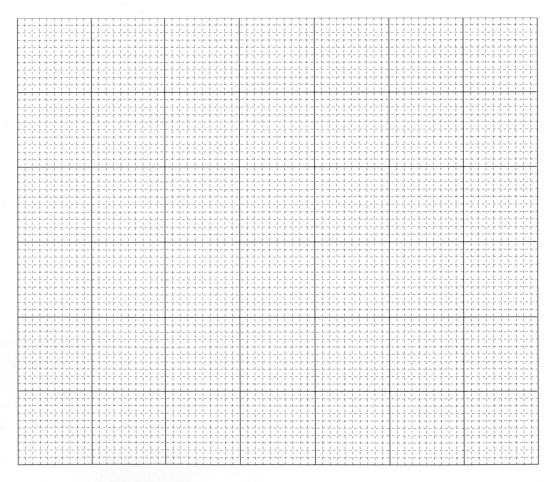

FIGURE 16-4. _____

Check your graph with your instructor before you continue.

Comprehension Check

1. In Family A, **family members 21 and 22** are expecting their ninth child. What is the probability that this child will have sickle-cell anemia? **Show your work!**

2. Closer to home, Michelle had a brother, Charles, who died of sickle-cell anemia. She is concerned about the chance of the condition appearing in her children. When blood samples were taken and placed under low-oxygen conditions, some of her red blood cells sickled. Those of her husband James, however, remained normal when tested. **Show your work and list the genotypes of all those mentioned in the problem.**

    Michelle _____   Charles _____   James _____

    What is the probability that Michelle's and James's children will have sickle-cell anemia? _____

    Sickle-cell trait? _____

Check your answers with your instructor before you continue.

Self Test

1. The ability to taste PTC is due to a **dominant allele (T)**. A woman nontaster married a man who was a taster. They had three children. Their two sons were tasters, but their daughter was a nontaster. All four grandparents were tasters.

    What are the genotypes of all the individuals mentioned?

    The woman _____   Her husband _____

    The two sons _____   The daughter _____

    Grandparents _____ _____

2. Your sister died from Tay-Sachs disease, inherited as a **recessive** allele **(d)**. You are married and planning to start your family. You are worried about the disease and decide to have genetic testing to see if you or your spouse is a carrier of the Tay-Sachs allele. The test results show that you are a carrier of the allele, but your spouse is not.

    What is the probability that you and your spouse will have a child with Tay-Sachs disease? **Show your work!**

3. Red-green color blindness is inherited through an **X-linked, recessive** allele **(b)**. Two parents, Fred and Ginger, have normal vision. They have two daughters, Takiyah and Kelly, who also have normal vision, and a color-blind son, David.

    Daughter Kelly has a color-blind son, Kevin. Daughter Takiyah has five sons, all with normal vision. What are the genotypes of all the individuals? **Show all your work!**

    Fred _____   Ginger _____   David _____

    Takiyah _____   Kelly _____   Kevin _____

    Takiyah's five sons _____

    If Kelly marries a man with normal vision, what is the probability that she will have

    a color-blind son? _____   a color-blind daughter? _____

4. Ralph has normal blood clotting, but he has two brothers and a sister who have hemophilia (an **X-linked, recessive** disorder). What are the most probable genotypes of Ralph's parents? **Explain your answer.**

5. Albinism in humans is expressed as the **absence** of pigment from the skin, hair, and eyes. Using the information in **Figure 16-5**, determine whether albinism is inherited as a **dominant or as a recessive** trait. **Affected** individuals are represented by **shaded** squares and circles.

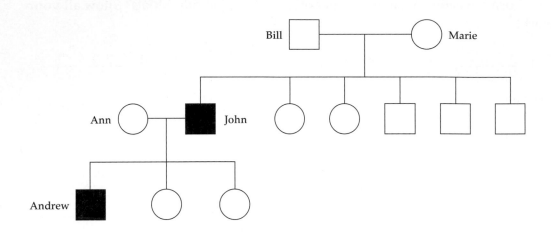

FIGURE 16-5.   Pedigree Showing Albino Individuals

a. Underneath **each circle and square** in the pedigree, **enter the genotype** of that individual.

b. **(Circle one answer).** Albinism is a trait that is probably inherited through a **dominant/recessive allele**.

Support your answer with information from the pedigree.

Introduction to Molecular Genetics

Objectives

After completing this exercise, you should be able to:

- demonstrate an understanding of DNA structure and the base-pairing rule
- explain the roles of all of the following in protein synthesis: DNA, mRNA, tRNA, and ribosomes
- complete the processes of transcription and translation from a strand of DNA and determine the sequence of amino acids in the resulting polypeptide chain
- explain how changes in the DNA code may affect protein synthesis and cause health problems.

CONTENT FOCUS

You have just spent several weeks thinking about genetics and the different ways in which traits can be transmitted from parents to children. Each chromosome is divided into sections called **genes** that are the basis of inheritance. The traits you inherited through your genes (such as hair color, blood type, presence of sickle-cell anemia, or ability to produce the hormone insulin) are all controlled by the production of proteins. **Genes contain the coded instructions your body uses to assemble the hundreds of different types of proteins that make you a unique individual.**

Genes are composed of a molecule known as **deoxyribonucleic acid (DNA)**. In this exercise, you will take a closer look at the **DNA molecule** and the role it plays in all the cells of your body.

ACTIVITY 1   REMOVING DNA FROM CELLS

> ### Note:
> Read these directions COMPLETELY before proceeding.

1. Work in groups. Get the following supplies: one piece of **calf thymus** (5 grams) in **a dish**, one graduated 10-ml **pipette with manual dispenser**, one **50-ml test tube**, a **test tube rack**, and a pair of **scissors**.

2. On your laboratory table, you will find a **blender, a piece of cheesecloth, a 500-ml graduated cylinder, a container of 6% saline solution, rubber bands, and a 600-ml beaker**.

    Using **scissors, mince** the piece of thymus **as much as possible** and place the pieces in the dish.

    Add 100 ml of 6% **saline solution** to the bowl.

> ### Note:
> All groups at your table will be using the blender at the same time, so you MUST coordinate your activities!

3. Coordinating with your partner groups, pour the saline containing the **minced thymus** from **each of the groups** at your laboratory table into the blender.

4. **Put the lid on the blender.** Blend the thymus mixture **on a low setting** for several minutes until no large pieces remain.

5. **Remove the lid** from the blender.

    Assemble **eight thicknesses of cheesecloth** (four pieces from the package, which is manufactured in double thickness). Place the pieces of cheesecloth over the top of the blender and **secure** the cloth **tightly** with the rubber band.

6. Making sure the cheesecloth is **tightly** fastened on the top of the blender, **pour the contents (filtered extract)** into the **600-ml beaker**.

> ### Note:
> At this point, you will go back into your ORIGINAL groups and complete the experiment.

7. Using a **graduated 10-ml pipette with manual dispenser**, remove 10 ml of **filtered extract** from the beaker and place the extract into a **50-ml test tube**. Place the test tube in the rack.

   Place your **used pipettes** in the waste container.

8. Get **one graduated 2-ml pipette** with **manual dispenser, one graduated 10-ml pipette**, and a **thin glass rod**.

   In your supply area, you will find a container of **10% SDS solution** (sodium dodecyl sulfate) and **an ice bucket containing 15-ml screw-cap tubes of 95% ethanol**.

9. To the **test tube containing the filtered thymus** extract, add **1 ml of SDS solution**. Tap the test tube several times to **gently** mix the contents.

Note:

Read the following directions COMPLETELY before proceeding.

---

10. Select a **tube of ethanol** from the ice bucket. Using a **clean**, graduated 10-ml pipette, add **10 ml of ice-cold 95% ethanol** to the test tube.

    With the test tube in the rack, place the filled pipette with its tip against the **inside wall** of the test tube. **SLOWLY** allow the ethanol to dribble down the inside of the tube, as **demonstrated in Figure 17-1. Don't shake the test tube during this procedure.**

pipette

FIGURE 17-1. Method for Adding Ethanol to Filtered Thymus Extract

11. The ethanol is lighter than the contents of the tube. When added according to directions, the ethanol will form a **clear layer ABOVE** the filtered thymus extract.

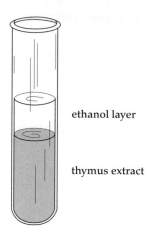

ethanol layer

thymus extract

FIGURE 17-2.   Ethanol Layered above Filtered Thymus Extract

12. Observe the test tube for **five minutes**. The DNA will gradually separate from the thymus mixture and rise into the ethanol layer.

    Describe the appearance of the DNA.

13. Let the DNA sample remain undisturbed while you complete the following **Comprehension Check** and **Activity 2**. When you have completed Activity 2, **observe** your DNA sample and see if any changes have occurred. **Record your observations** below.

14. To remove the accumulated DNA from the test tube, follow the directions for **DNA spooling as** below:

   a. **Gently insert** the glass rod through **the ethanol layer** into the **accumulated DNA**.

   b. **Carefully twirl** the rod between your fingers, winding the DNA onto the rod, imitating thread on a spool.

   c. **Slowly remove the rod** from the test tube.

✓ Comprehension Check

1. The thymus is an organ made up of many different types of cells. What percentage of thymus cells contain DNA?

   a. 0%   d. 75%
   b. 25%   e. 100%
   c. 50%

   Explain your answer.

2. If you removed DNA from a sample of human thymus, how many separate DNA molecules would be removed from **each cell**? _____

3. **(Circle one answer.)** If you repeated the same experiment with an equal number of kidney cells, the amount of DNA collected would **increase/decrease/stay the same**.

    Explain your answer.

4. **(Circle one answer.)** If you repeated the same experiment with an equal number of sperm cells, the amount of DNA collected would **increase/decrease/stay the same**.

    Explain your answer.

5. *Challenge Question!* **(Circle one answer.)** If you repeated the same experiment with an equal number of red blood cells, the amount of DNA collected would **increase/decrease/stay the same**.

    Explain your answer.

Check your answers with your instructor before you continue.

ACTIVITY 2   THE BASICS OF DNA STRUCTURE

Preparation

The extraction of DNA that you just performed showed that DNA is present in cells. It does not, however, give you much information about its actual structure.

DNA molecules are composed of small building blocks called **nucleotides**.

1. Each DNA nucleotide is composed of three smaller molecules hooked together:

    one five-carbon sugar (deoxyribose)
    one phosphate
    one nitrogen base

2. Four different types of nucleotides are needed to build a DNA molecule.

    Each of these four nucleotides has a **different nitrogen base: adenine, guanine, cytosine, or thymine.**

3. DNA has a structure similar to a ladder. The two sides of the ladder are composed of **alternating sugar and phosphate molecules.**

4. Each sugar molecule is attached to **one nitrogen base**. The two strands of DNA are attached by bonds between the nitrogen bases on each side of the ladder.

    Each nucleotide base only bonds with **one specific partner**. The combination of two bases is called a **base pair**.

    Adenine always bonds with thymine. A-T

    Guanine always bonds with cytosine. G-C

5. Fill in the blanks on the incomplete DNA molecule in **Figure 17-3**.

    Use the symbol **"P"** for **phosphate, "S"** for **sugar**, and **"A," "T," "C," or "G"** for the **appropriate nitrogen bases.**

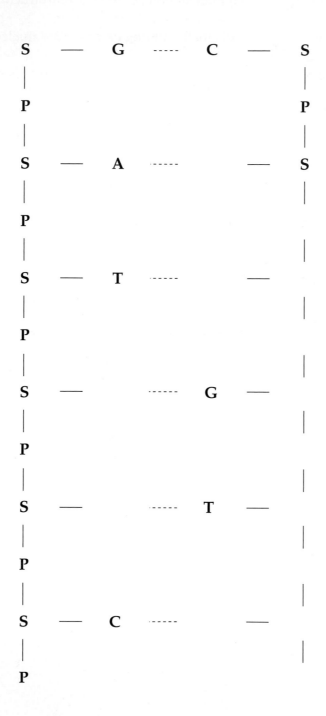

FIGURE 17-3. Incomplete DNA Molecule

✓ Comprehension Check

1. On **Figure 17-3**, draw a box around one complete nucleotide.

2. **How many nucleotides** are shown in the DNA molecule in **Figure 17-3**? _____

3. How many **different** types of nucleotides were used to construct the DNA molecule in **Figure 17-3**? _____

Check your answers with your instructor before you continue.

ACTIVITY 3   BUILDING A MODEL OF DNA

1. Work in groups. Get a **DNA-model kit**.

2. Construct the **support stand for the model** by assembling:

    one gray tube (eight inches long)

    three green tubes (two inches long)

    one black connector with four prongs

    Set the stand aside for later use.

3.   Separate the parts of the DNA model according to the descriptions in **Table 17-1**.

TABLE 17-1 PARTS OF THE DNA MODEL	
PARTS OF THE DNA MOLECULE	DESCRIPTION OF MODEL PART
Deoxyribose sugar	Black, three prongs
Phosphate	Red, two prongs
Sugar-to-phosphate connectors	Yellow tube
Adenine (A) base	Blue tube
Thymine (T) base	Red tube
Guanine (G) base	Green tube
Cytosine (C) base	Gray tube
Base-to-base connectors	White, two prongs

4.   Using your knowledge of DNA structure, **assemble the DNA model**. Use the same **base-pairing sequence** as that shown in **Figure 17-3**.

5.   Place the model onto the stand by inserting the vertical tube on the stand through the holes in the **white base-to-base** connectors.

     Twist the model as you slide it onto the stand. Stop twisting when the entire molecule fits onto the stand.

     The twisted shape of the DNA molecule is known as a **double helix**.

✓ Comprehension Check

1. How many nucleotides are present in your DNA model? _____

2. What is the base-pairing rule?

3. Why would it be appropriate to call a DNA molecule a polynucleotide?

4. Assume that the model you just built is an exact representation of **your** DNA code.

   a. Would you use the same bases to construct your lab partner's DNA? _____

   b. Would you assemble the bases in the same order to make a model of your lab partner's DNA? _____

   Explain your answer.

Check your answers with your instructor before you continue.

Each gene has the instructions to make a single polypeptide chain. These instructions are part of the genetic code. Polypeptide chains are the structural units of proteins. A polypeptide chain is made of many amino acids hooked together.

The key to the genetic code is the sequence of nitrogen bases along **one side** of the DNA molecule. To construct a protein, you must know the **order of the bases**. The code is written in **three-letter "words."** Each of these words (called **triplets**) tells the cell which amino acid should come next when building a protein.

For proteins to function properly, the amino acids must be assembled in the correct order.

5.   How many **triplets** are present along one side of your DNA model? _____

6.   How many **amino acids** will be present in the protein made from your model? _____

Note:

DON'T take your DNA model apart yet.

ACTIVITY 4   STEPS OF PROTEIN SYNTHESIS

Preparation

When a particular protein is needed by the body, regions of the double helix unwind so that a cell gains access to the genes that contain the coded information to make that protein. Protein synthesis has two steps: **transcription** takes place in the nucleus and **translation** occurs in the cytoplasm. Both steps require molecules of **RNA (ribonucleic acid)**.

Although the **nucleus** contains the **instructions** for protein synthesis, the **machinery to make proteins** is located in the **cytoplasm**. The coded information is **transferred** from the nucleus to the cytoplasm during **transcription**.

Transcription

1.   During transcription, DNA bases are **copied** to form a single strand of RNA, called **messenger RNA (mRNA)**. As with the DNA, mRNA is divided into coded three-letter words. In mRNA, these words are called **codons**.

2. The **base-pairing rule** is used to form messenger RNA **with one exception**. RNA molecules do not have the nitrogen base thymine. They have **uracil** instead.

Base-pairing to form mRNA:

DNA BASE	MRNA BASES
C	G
G	C
T	A
A	U

3. The coded information to make a protein appears along **one side** of the double helix. Practice transcription by filling in the correct messenger RNA codons in **Table 17-2**.

4. How many **amino acids** would be in this protein? _____

TABLE 17-2
TRANSCRIPTION OF MRNA

DNA triplets	CGC	ATA	GAC	TTT	CTT	ACT	TAG	CAT	AAA
mRNA codons									

5. Which of the five types of nitrogen bases is **not** found in mRNA? _____

Translation

1. A cell needs **amino acids** to construct proteins. The amino acids are carried to the ribosomes by another type of RNA molecule, called **transfer RNA (tRNA)**.

A tRNA molecule has **two functional ends**.

One end picks up amino acids in the cytoplasm (see **Figure 17-4**).

2. The other end is called the **anticodon**. It contains **three nitrogen bases** that can form a base pair with a **matching codon** in the messenger RNA.

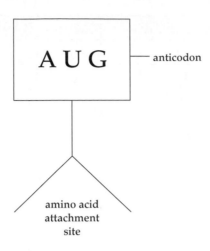

FIGURE 17-4. Structure of a Transfer RNA Molecule

3. Each type of tRNA can carry **only one type of amino acid**.

There are enough different types of transfer RNA molecules to carry all the different types of amino acids needed to make your body's proteins.

4. **Where do the transfer RNA molecules take the amino acids?**

They take them to **ribosomes**, organelles in the cytoplasm where proteins are manufactured.

Ribosomes are made of proteins and a third type of RNA called **ribosomal RNA**.

5. **Ribosomes read messenger RNA codons and accept amino acids brought by transfer RNA molecules**.

Ribosomes hook amino acids together in the **order specified by the messenger RNA** codons to construct the polypeptide chain.

Summary of Protein Synthesis

■ DNA contains the instructions to make polypeptide chains. A region of the DNA double helix unwinds. Coded instructions to make a protein are exposed.

■ This information is carried to the cytoplasm by messenger RNA molecules (transcription).

■ Amino acids in the cytoplasm are used to build polypeptides.

■ Transfer RNA molecules pick up the amino acids and transport them to ribosomes, the locations where proteins are made.

■ Ribosomes bond amino acids together according to the instructions in the genetic code.

ACTIVITY 5   BUILDING A REAL PROTEIN

Preparation

Imagine the following situation: you are about to give birth to a baby. The brain produces the hormone **oxytocin** (a small protein), which causes uterine muscles to contract for childbirth. After birth, this same hormone causes muscles in the mammary glands to contract, releasing milk for nursing the baby.

   Suppose this is your first baby. How does the brain know how to manufacture oxytocin if it has never been needed before? The information is stored in your DNA *"reference"* library.

1. Using the steps outlined in **Activity 4** as guidelines, build the protein **oxytocin**.

   Oxytocin is one of the smallest proteins with only nine amino acids connected into a single polypeptide chain.

2. **Transcribe** the DNA triplets that code for oxytocin into messenger RNA codons and add them to the appropriate spaces below.

DNA TRIPLETS	ATG	TAT	GTT	TTG	ACG	GGA	GAC	CCC

mRNA CODONS								

3. The messenger RNA must detach from the DNA and leave the nucleus for **translation** to take place.

   Cut out the strip of mRNA and move it to the ribosome in **Figure 17-5**. Notice that the ribosome has two parts, separated by a groove. The mRNA should be placed along the groove between the upper and the lower sections.

4. Each transfer RNA molecule in **Figure 17-6** is ready to deliver an amino acid to the ribosome.

   Where should each transfer RNA molecule deliver its amino acid?

   Cut out the **transfer RNA** molecules.

   Place each tRNA molecule under the correct mRNA codon.

Hint:

You will be able to match each tRNA with its codon by following the base-pairing rule.

---

5. Refer to **Table 17-3** to determine the name of each amino acid in your oxytocin polypeptide chain.

   Each box at the bottom of **Figure 17-5** represents **one amino acid** in the chain.

   Write the names of the amino acids in order in each box.

amino acids that make up the oxytocin polypeptide chain

FIGURE 17-5. Translation

Check your oxytocin molecule with your instructor before you continue.

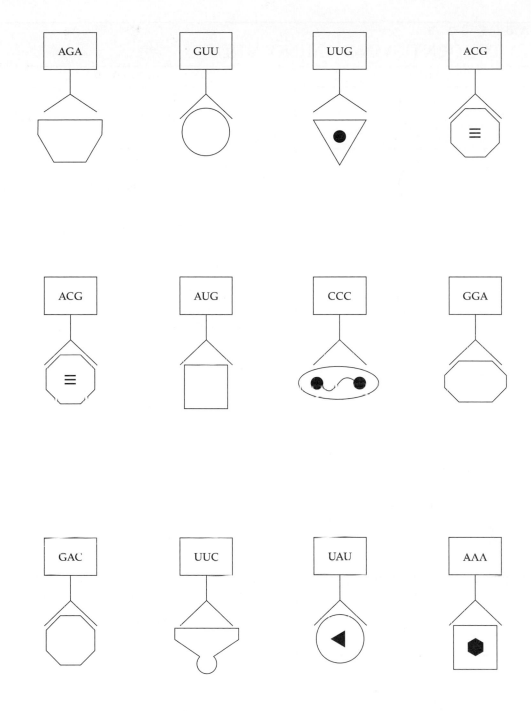

FIGURE 17-6. Molecules of tRNA

TABLE 17-3

KEY TO IDENTIFYING AMINO ACIDS

SHAPE	NAME OF AMINO ACID
	Histidine
	Glutamine
	Asparagine
	Cysteine
	Tyrosine
	Glycine
	Proline
	Leucine
	Alanine
	Isoleucine
	Serine

Comprehension Check

1. If the second amino acid in your polypeptide chain were **valine**, would the protein still be oxytocin? _____ **Explain your answer**.

2. **(Circle one answer.)** The DNA triplet A A A would be transcribed into the mRNA codon **T T T/U U U**.

3. Put the following steps of protein synthesis in the correct order.

    _____ tRNA molecules pick up amino acids

    _____ mRNA transcribed

    _____ DNA double helix unwinds

    _____ mRNA binds to ribosome

    _____ Ribosome bonds amino acids together

    _____ tRNA anticodon links with mRNA codon

    _____ mRNA leaves nucleus

    _____ Polypeptide chain completed

Check your answers with your instructor before you continue.

Self Test

Each DNA triplet represents one letter of the alphabet. **1)** Translate the DNA triplets into mRNA codons. **2)** Use the charts of mRNA codons to find the correct amino acids. **3)** Write the alphabet letter for that amino acid above the appropriate DNA triplet to decode the message.

TAT' ATC CTT   CCC CAG T'GA TAT TGA—TGC GTG CTT

CCC CTT TTG CTT TGA TAT ACG   ACG CAG CTA CTT!

ACC CGA TAT TGA   CGA TAC TAT TTG ACT TGA CTT,

ACC GTG CAG   CGA GAG GGG GTG CGA-

GTT CTT TGA TAT TTT CTT CTA   TGA GTG TAT AGA?!

1ST LETTERS	2ND LETTERS	3RD LETTERS	AMINO ACID	LETTERS TO DECODE DOCUMENT
A	A	A	lysine	Z
		C	asparagine	N
		U	asparagine	N
		G	lysine	Z
	C	A	threonine	T
		C	threonine	T
		U	threonine	T
		G	threonine	T
	G	A	arginine	R
		C	serine	C
		U	serine	C
		G	arginine	R
	U	A	isoleucine	I
		C	isoleucine	I
		U	isoleucine	I
		G	methionine/ start	M

1ST LETTERS	2ND LETTERS	3RD LETTERS	AMINO ACID	LETTERS TO DECODE DOCUMENT
U	A	A	terminator 1	J
		C	tyrosine	Y
		U	tyrosine	Y
		G	terminator 2	V
	C	A	serine	S
		C	serine	S
		U	serine	S
		G	serine	S
	G	A	terminator 3	U
		C	cysteine	C
		U	cysteine	C
		G	tryptophan	W
	U	A	leucine	L
		C	phenylalanine	F
		U	leucine	L
		G	leucine	L

		A	glutamine	B
C	A	C	histidine	H
		U	histidine	H
		G	glutamine	B
	C	A	proline	P
		C	proline	P
		U	proline	P
		G	proline	P
	G	A	arginine	R
		C	arginine	R
		U	arginine	R
		G	arginine	R
	U	A	leucine	L
		C	leucine	L
		U	leucine	L
		G	leucine	L

		A	glutamic acid	E
G	A	C	aspartic acid	D
		U	aspartic acid	D
		G	glutamic acid	E
	C	A	alanine	A
		C	alanine	A
		U	alanine	A
		G	alanine	A
	G	A	glycine	G
		C	glycine	G
		U	glycine	G
		G	glycine	G
	U	A	valine	O
		C	valine	O
		U	valine	O
		G	valine	O

Biotechnology: DNA Analysis

Objectives

After completing this exercise, you should be able to:

- discuss the differences between genes and non-coding regions of DNA
- summarize how the process of gel electrophoresis separates DNA molecules
- explain the role of non-coding DNA regions in producing a DNA profile
- draw conclusions that are supported by analysis of RFLP and STR profiles
- interpret DNA data presented in the form of tables, charts, and/or graphs
- explain how STR analysis can be used to determine paternity
- apply your knowledge of DNA fingerprinting to real-life situations.

CONTENT FOCUS

Half of your 46 chromosomes were inherited from your mother and the other half from your father. Consequently, **everyone has a unique DNA pattern** (except for identical twins). Tissue samples containing DNA can be used for identification in criminal cases, in paternity suits, and in cases where visual identification is not possible. The technique used to make these genetic comparisons produces a **DNA profile (also called a DNA fingerprint).**

DNA for examination is isolated with a process similar to the method you used in the DNA spooling laboratory. As you know, DNA is located in the nucleus of all body cells. Therefore, DNA can be extracted from any tissue. Common sources include **white blood cells, hair roots, semen, saliva** (which contains **epithelial cells**), and other body tissues.

So far, you have studied basic genetics and the structure of the DNA molecule. You have seen that various traits are produced on the basis of your genetic code (DNA). Not all of a person's DNA codes for synthesis of proteins, however. **DNA molecules also have areas that are not genes**.

Originally, this non-coding DNA was called "**junk DNA**" because scientists did not understand or recognize the function for these sections of the chromosomes (see **Figure 18-1**). They originally believed that these sequences were present to simply fill the gaps between the genes. Although the functions for most non-coding regions have not yet been discovered, scientists are beginning to find evidence that some of these repetitive DNA sequences play important roles in cellular metabolism and inherited diseases.

Regions of non-coding DNA vary a great deal among individuals, and when specific DNA regions are studied, scientists can use this information to establish human identity, analyze evolutionary trends, and determine predispositions to certain diseases.

Gene 1   Junk DNA   Gene 2

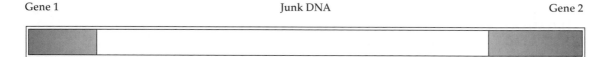

FIGURE 18-1.   Position of Genes and Non-coding Regions of DNA

The pattern of non-coding regions between any given two genes is unique and always has the same repeating pattern of nucleotides (for example **C A T**). The number of times these nucleotides are repeated, however, is highly variable among people.

As you can see in **Figure 18-2**, Person A has nine repeating **C A T** segments, while Person B has only five. Some people have dozens of repeats of the same pattern.

Each person has a distinct **DNA profile** (or fingerprint).

One of the original processes used to create a DNA profile is called **RFLP Analysis**. RFLP stands for **restriction fragment length polymorphism**. DNA extracted from a blood or a tissue sample is cut into small pieces at specific locations using special chemicals called **restriction enzymes**.

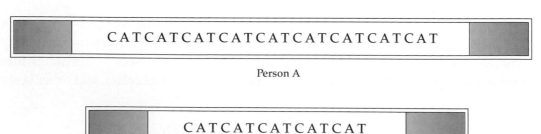

CATCATCATCATCATCATCATCATCAT

Person A

CATCATCATCATCAT

Person B

FIGURE 18-2.   Repeating **Non-Coding** DNA Segments

Fragment length polymorphism refers to the fact that the **pieces (fragments) of non-coding regions are different lengths for different people**.

To make a reliable "RFLP match," molecular geneticists use several different non-coding DNA regions. When several different regions are compared, the odds against a mistaken match can be **more than a billion to one**.

ACTIVITY 1   RFLP FINGERPRINTS

Preparation

An early form of forensic DNA analysis used **restriction enzymes (enzymes that cut the DNA molecule at specific sites)** to produce fragments of DNA. These fragments vary in size from individual to individual based on the number of times that a core sequence is repeated.

Fragments of DNA are separated using a process called **gel electrophoresis**. The **gel** is a thin sheet of gelatin supported by a glass plate with **electrodes** attached at both ends. One end of the gel has **a positive** charge and the other end has a **negative** charge.

The DNA fragments are attracted to the **positive end** of the gel plate. **Smaller, lighter** fragments migrate more easily through the gel and therefore travel farther in a given time period than larger, **heavier** fragments. The light and heavy fragments form bands across the gel layer (see **Figure 18-3**).

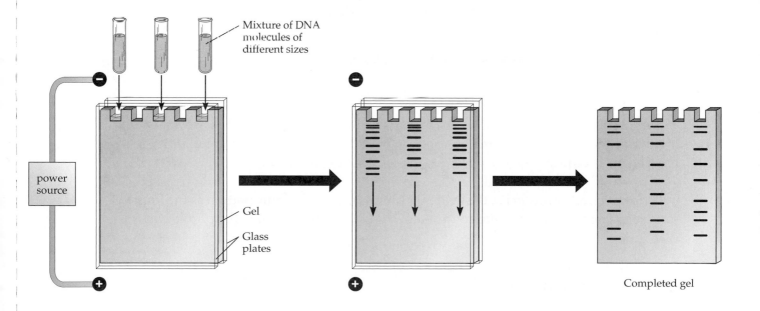

FIGURE 18-3. Gel Electrophoresis of DNA

The DNA bands of interest are **marked** with special **molecular probes**. Probes are small pieces of DNA that use the **base-pairing rule** to locate and bind **only** to the fragments that will be used to form the DNA profile.

The sheet of DNA bands is "photographed" with **X-ray film**. The **only** bands visible on the developed film are those that have been **labeled with molecular probes**. The X-ray **pattern of bands** for several DNA locations is analyzed to create a **DNA profile** for an individual (see **Figure 18-4**, which contains samples of several DNA profiles). Each profile is represented by a **vertical column of DNA bands**.

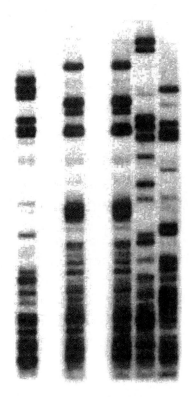

FIGURE 18-4. Sample DNA Fingerprints

For RFLP testing, **several different regions (called loci) of junk DNA are analyzed**. If **ANY** of the loci **do not match** between an evidence sample and a known individual, **the person is eliminated** as a possible source of the evidence DNA.

If, however, **all of the loci in the evidence sample** exhibit the **same pattern as the known** sample, calculations are conducted to determine how rare or common the profile was.

Comprehension Check

Figure 18-5 shows some samples of DNA profiles produced by multiple tests of RFLP loci. Each profile is represented by a **vertical column of DNA bands**.

Two profiles match and two do **not** match. Matching profiles must have **exactly** the same banding patterns. The bands have been numbered for reference.

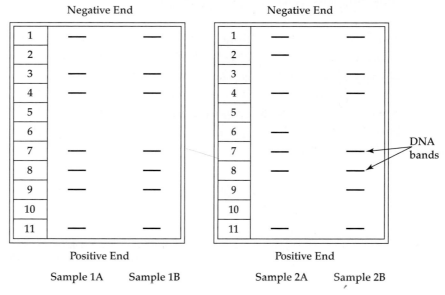

FIGURE 18-5. Four DNA Profiles

1. (**Circle one answer.**) The DNA sample was placed in the gel at the **negative/positive** end of the tray.

2. List the bands that **don't match** in samples **2A and 2B**. _____

3. Which band in each sample contains the **heaviest** DNA fragments? _____

4. Which band contains the **lightest** fragments? _____

5. How many DNA bands are present in **sample 2B**? _____

6. (**Circle one answer.**) If the DNA fragments are migrating toward the positive end of the gel, DNA molecules must have a **negative/positive** charge.

7. *Challenge Question!* What could you conclude from the RFLP analysis if you knew that the DNA in **sample 1A** was from blood and the DNA in **sample 1B** was from skin?

Check your answers with your instructor before you continue.

ACTIVITY 2   SOLVING A CRIME USING RFLP FINGERPRINTS

Preparation

On January 1, at about 2:00 A.M., the police responded to a report of gunshots at the 600-block of Knorr Street. Arriving at the scene, the officers found a parked Dodge pickup. In the front seat was a deceased male with several gunshot wounds.

Shoe marks were visible in a pool of blood outside the truck, but they were not clear enough to identify the type of footwear that made them. Homicide detectives have narrowed the field to three suspects.

Suspect A was arrested near the scene. He was wearing black work boots that appeared to have dried blood on the soles.

At the home of **Suspect B**, investigators recovered a pair of tennis shoes that also appear to have dried blood on the soles. Suspect B says she had a severe nosebleed, which might have resulted in blood on her shoes.

Suspect C was the roommate of the deceased. When officers went to the apartment to notify him of the death, Suspect C was packing a suitcase. He was wearing high-top sneakers that appeared to have dried blood on the soles and sides. When questioned about the blood and the murder, Suspect C had no comment.

All the samples were tested and identified as human blood. The Homicide detectives have asked the Central Police DNA Laboratory to analyze the blood evidence using RFLP analysis. You are the forensic scientist assigned to the case. **Figure 18-6** shows the DNA profiles from all the blood samples.

1. Note that the box under the **first column** in **Figure 18-6** (blood of victim) contains the number **1**. This is the first unique DNA profile found.

2. Moving from left to right, examine each DNA profile.

    If the new profile **does not match any previously examined profile**, enter a **new number** in the box under the column (**2, 3, 4,** etc.).

    If the new DNA profile **is an exact match** of a previous profile, enter the **number you gave to the profile that matches this new profile**.

    Continue with this process until you have examined all the DNA profiles in **Figure 18-6**.

	Blood of Victim	Blood on Ground at Crime Scene	Blood of Suspect A	Blood of Suspect B	Blood of Suspect C	Blood from Suspect A's Shoes	Blood from Suspect B's Shoes	Blood from Suspect C's Shoes
1			—		—	—		
2	—	—						—
3	—	—			—			—
4			—			—		
5			—	—		—	—	
6				—	—			
7	—	—		—			—	—
8					—			
9			—		—	—		
10	—	—						—
11				—			—	
12			—		—	—		
13				—			—	
14								
15	—	—						—
16	—	—						—
17								
18				—			—	
19								
20	—	—	—			—		
	1							

FIGURE 18-6. DNA Profiles—Knorr Street Homicide

✔ Comprehension Check

1. How many **unique** profiles were found? _____

2. Did any profiles match the blood from **Suspect A's shoes**? _____

    If so, list them:

3. Did any profiles match the blood from **Suspect B's shoes**? _____

    If so, list them:

4. Did any profiles match the blood from **Suspect C's shoes**? _____

    If so, list them:

5. Did any profiles match the **blood from the deceased male**? _____

    If so, list them:

6. Did any profiles match the **blood found on the ground** outside the truck? _____

    If so, list them:

7. Do these **results** make any individual a more likely suspect than the others? _____

    Explain your answer, citing facts from the DNA analysis.

8. What other types of evidence would be helpful to be sure you had found the murderer?

 Check your answers with your instructor before you continue.

ACTIVITY 3 SHORT TANDEM REPEAT (STR) ANALYSIS

Frequently, the amount of DNA collected at crime scenes is insufficient for RFLP analysis. For this reason, forensic scientists began to use a procedure called the **polymerase chain reaction (PCR)**. PCR allows the scientist to **target specific chromosome locations and make copies of the DNA**. By using this technique, scientists can copy samples containing a small amount of DNA (like the back of a licked envelope).

Scientists currently use non-coding **DNA loci** called **short tandem repeats (STRs)** in forensic DNA testing. **An STR is a small sequence of bases that is repeated.** The number of times that the unit is repeated varies from person to person. For example, an STR with the **base sequence** AATC may be repeated between 12 and 24 times.

As with all of your DNA, the alleles for this example STR locus would be inherited from your parents.

Each individual would have **two alleles for this STR locus** (recall that each person inherits one allele from each parent).

So, an individual's **genotype for this locus could be 13, 14 (meaning that the allele this person inherited from one parent repeats the STR base sequence 13 times and the other allele repeats 14 times).**

To make an **STR profile**, the DNA sample is tagged with a **fluorescent label**. The label is detected by a **laser**, which sends a **signal to a computer**. This signal produces a peak on a computer printout.

The pattern of peaks on an STR profile (see **Figure 18-7**) represents the alleles at each locus (recall that there are two alleles at each locus). If you see only **one peak** for a locus, the individual has **two alleles that are alike. They are homozygous** at that locus.

In the sample locus to the right, note that there are **numbers below each peak**.

These two numbers are used to designate the person's **genotype**.

In this example, the **genotype is 13,14.**

The 13 allele has the STR "phrase" AATC repeated 13 times. The 14 allele has the STR phrase AATC repeated 14 times.

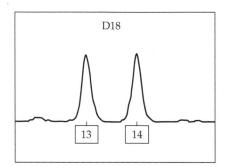

The **letter and number above the peaks** are **abbreviations** that represent the different STR loci.

FIGURE 18-7.   Computer Printout of Sample Alleles at One STR Locus

As with RFLP, an STR profile does not consist of only one locus. **Multiple loci are used to develop a DNA profile.** The forensic community has adopted **13 core STR loci** that are tested by all forensic laboratories. By having all labs perform the same test, results from around the country can be compared. These results are currently being stored and compared in a **computer database system** called **CODIS (Combined DNA Index System.)**

DNA samples are frequently analyzed in **two STR sets. Group 1** contains **nine STR loci** plus one locus, abbreviated **AML,** that is used for **gender identification (XX or XY).**

Group 2 contains **the remaining four STR loci, plus the gender locus, plus two STR loci from Group 1 that are repeated** as a **control in Group 2**, to ensure that samples originated from the same individual.

A complete STR profile is shown in **Figure 18-8.**

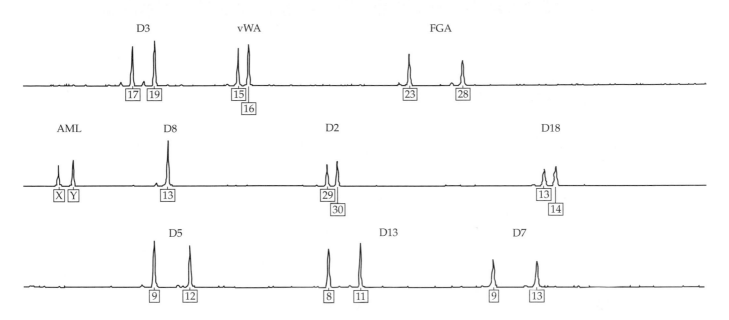

Group 1

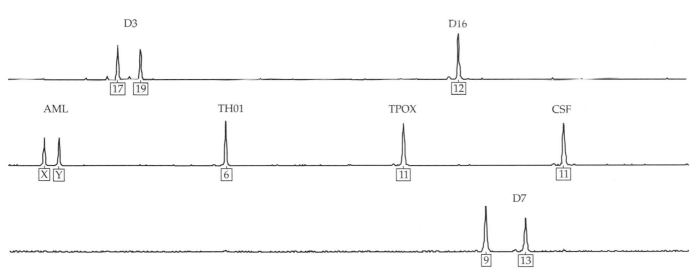

Group 2

FIGURE 18-8. Example of an STR Profile.

✔ Comprehension Check

1. **(Circle one answer.)** This DNA profile was taken from a **male/female. Explain your answer.**

2. Using the STR abbreviations on the profile, which two loci are the controls (repeated in both Groups 1 and 2)?

3. How many alleles in **Group 1** of the STR profile are homozygous? _____

   How many are heterozygous? _____

   How many alleles in **Group 2** of the STR profile are homozygous? _____

   How many are heterozygous? _____

   Explain your answers.

4. What do the numbers 9 and 13 designate in the **D7 locus** of Group 2?

5. *Challenge Question!* From your answer to Question 3, can you tell if any of these alleles are dominant or recessive? **Explain your answer.**

ACTIVITY 4   A CRIMINAL CASE USING STR EVIDENCE

The local police department responded to a call regarding a sexual assault. Upon arriving at the scene, they found the victim lying on the floor, partially clothed, and crying. The victim, Jane Doe, was transported to the hospital for treatment and evidence collection. Using a brief description of the suspect, the police apprehended two individuals, but the victim was not able to make a positive identification.

At the hospital, **two evidence samples** were collected from the body of the victim. The evidence samples **tested positive for semen** and were sent to the lab for STR analysis. In addition, **known blood samples from all individuals** were obtained and submitted for STR analysis.

1. Interpret the STR profiles in **Figures 18-9 through 18-13** on the following pages and **write the genotypes in the STR table** below.

Grp 1 Sample	D3	v WA	FGA	AML	D8	D2	D18	D5	D13	D7
Semen sample one										
Semen sample two										
Victim										
Suspect one										
Suspect two										

Grp 2 Sample	D3	D16	AML	TH01	TPOX	CSF	D7
Semen sample one							
Semen sample two							
Victim							
Suspect one							
Suspect two							

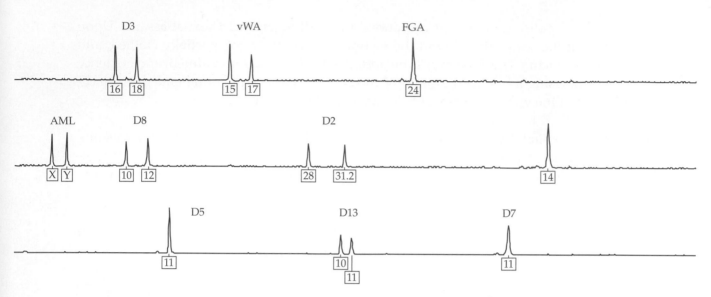

Group 1

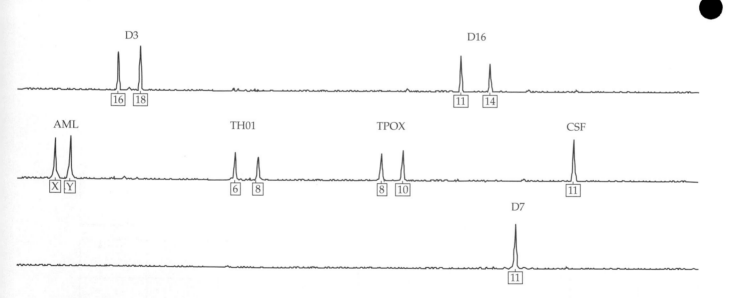

Group 2

FIGURE 18-9. STR Profile of Semen Sample One

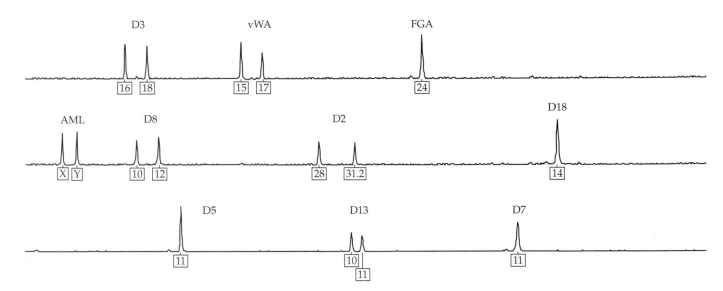

Group 1

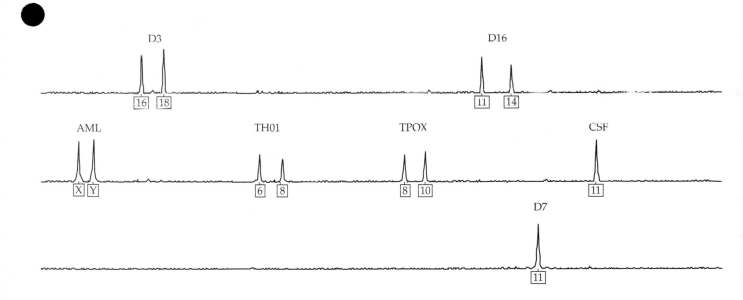

Group 2

FIGURE 18-10. STR Profile of Semen Sample Two

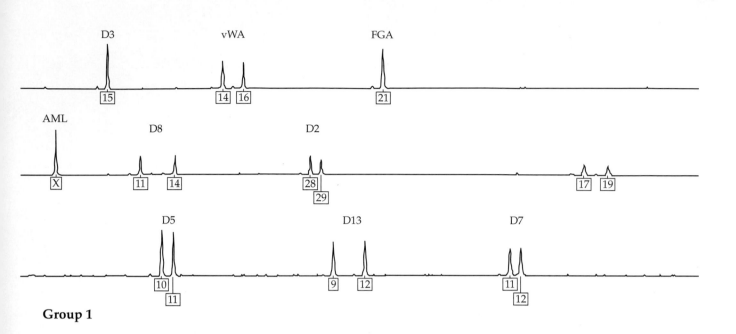

Group 1

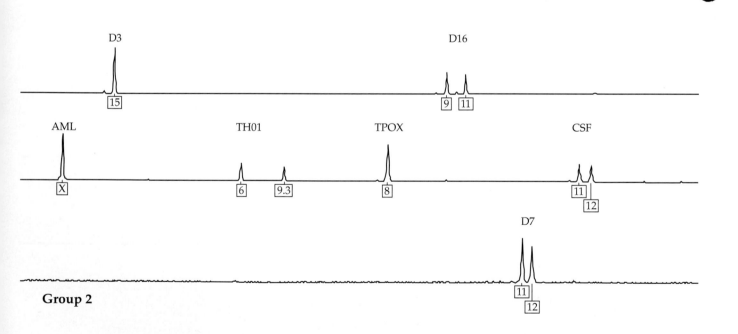

Group 2

FIGURE 18-11. STR Profile of Known Blood Sample from the Victim

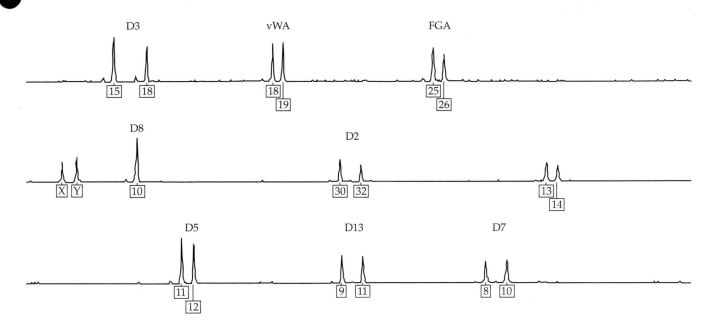

Group 1

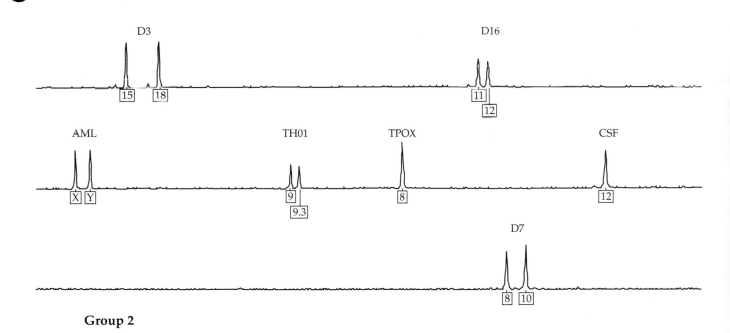

Group 2

FIGURE 18-12. STR Profile of Known Blood Sample from Suspect One

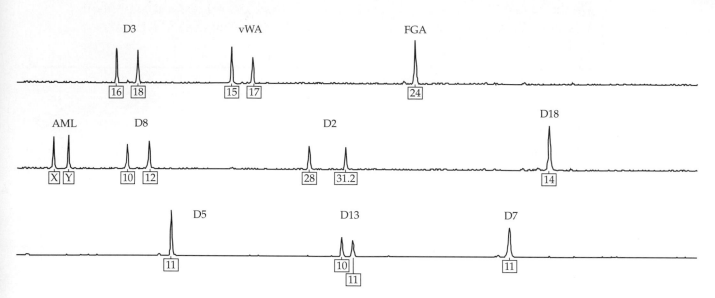

Group 1

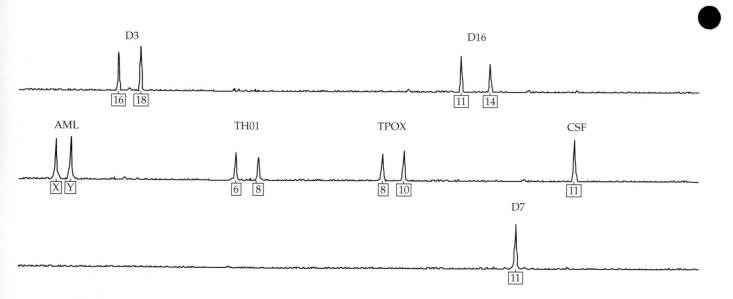

Group 2

FIGURE 18-13. STR Profile of Known Blood Sample from Suspect Two

✔ Comprehension Check

1. On the basis of the results of your analysis, what conclusions can you draw about this case? **Be specific. Support your answer with facts** from the STR profiles.

2. The detective investigating this case asks you the following question. Please respond.

   "If the DNA profile from the evidence matches the suspect, he must be guilty of sexual assault, right?"

3. **(Choose one answer.)** When a blood sample is submitted for STR analysis, the DNA tested is taken from **red blood cells/white blood cells/platelets. Explain your answer**.

4. What results would you expect from a comparison of STR profiles taken from the following? **Explain your answers**.

    a.   Identical twins:

    b.   Fraternal twins:

Check your answers with your instructor before you continue.

ACTIVITY 5   A PATERNITY SUIT INVOLVING STR ANALYSIS

Preparation

STR profiles can also be used to help identify a child's biological parents. As you recall from studying meiosis, a child inherits half of its genetic information from each parent. This makes matching of DNA profiles more challenging.

If a child has alleles in his or her STR profile that don't match those of the mother, these alleles must have been inherited from the father.

A young woman claims that her two-year-old son is the child of one of the members of the hit rock group "First Impressions." During the time of conception, she was dating all four of the band members.

In her sworn testimony, the mother states, "The father is one of the band members. They were the people I spent time with during that whole year. The problem is, I don't know which musician is the biological father of my baby."

None of the members of the band have admitted to fathering the child. During the investigation, all members of the rock group agreed to donate samples of their blood for DNA analysis. You are the forensic scientist appointed by the court to analyze the DNA evidence.

The profiles for all parties involved are summarized in the following STR table.

1. Compare the child's STR profile with the mother's. In the child's profile, **cross out** any **Group 1 and Group 2** alleles that are the same as the mother's (as shown in the example on page 364).

2. Next, compare the alleles of each of the potential fathers with the child's profile. **Circle all matching alleles.**

3. On the basis of your analysis of the STR profiles, which of the men could be the father of the child? **Explain** your answer, using facts from the STR analysis.

Check your answers with your instructor before you continue.

Grp 1	D3	vWA	FGA	AML	D8	D2	D18	D5	D13	D7
Child	13, **18**	12, 12	20, 25	X, X	9, 18	28, 30	10, 12	7, 16	13, 13	6, 9
Jenny Smith	12, **18**	12, 15	20, 20	X, X	9, 18	28, 32	12, 19	12, 16	13, 14	6, 6
Peter Niles	13, 13	11, 12	22, 25	X, Y	9, 9	31, 32	9, 10	10, 12	8, 9	9, 11
Phillip Cruze	12, 13	12, 20	25, 29	X, Y	10, 18	30, 30	10, 18	7, 9	11, 13	9, 15
Mike Miller	13, 19	12, 12	24, 26	X, Y	9, 15	31, 31	14, 15	13, 16	15, 15	6, 8

Grp 2	D3	D16	AML	TH01	TPOX	CSF	D7
Child	13, **18**	11, 14	X, X	9, 11	7, 11	13, 14	6, 9
Jenny Smith	12, **18**	8, 11	X, X	10, 11	7, 13	6, 14	6, 6
Peter Niles	13, 13	9, 13	X, Y	5, 7	8, 9	12, 14	9, 11
Phillip Cruze	12, 13	14, 16	X, Y	9, 10	11, 13	11, 13	9, 15
Mike Miller	13, 19	11, 16	X, Y	7, 11	8, 12	9, 10	6, 8

Self Test

1. What is meant by "non-coding" DNA regions?

2. How can non-coding DNA regions be used to determine someone's identity? **Be specific**.

3. How is comparing STR profiles to determine paternity different from trying to match a suspect to a blood sample?

4.   You are a park ranger at Yellowstone National Park. You have discovered the internal organs of a deer just inside the park boundaries. Because the park is protected from hunting, you notify the local police to be on the alert for a recently killed deer. The police spot a truck carrying the carcass of a large deer a few miles outside the park. They suspect this is the deer that was killed inside the park boundaries, but they need some evidence linking this deer to the remains found in the park.

What test(s) could determine whether the deer in the truck was killed inside Yellowstone Park? **Explain your answer**.

5.   On Saturday, September 28, a young boy was bitten by a large black dog on the corner of Fifth Avenue. Investigating the situation further, police officers learned that a neighbor has a large black dog that often barks ferociously at people walking down the street. The owner maintained that the dog made a lot of noise, but was basically friendly. The officers, however, discovered a substance, which appeared to be blood, on the dog's collar. The owner states that he cut his hand on the dog collar and that the blood is his.

The boy's family decide to sue the dog's owner. They wish to be compensated financially and also have the dog destroyed. However, the dog's owner is persistent that his dog never left his fenced yard on the date in question. You are a forensic scientist for a private firm that has been asked to evaluate the case for the civil trial. All samples have been tested and found to be human blood. Interpret the results of the DNA testing as presented in **Figures 18-14** through **18-16**.

Summarize your findings in the dog bite investigation. **Support your statements with evidence** from the DNA fingerprints.

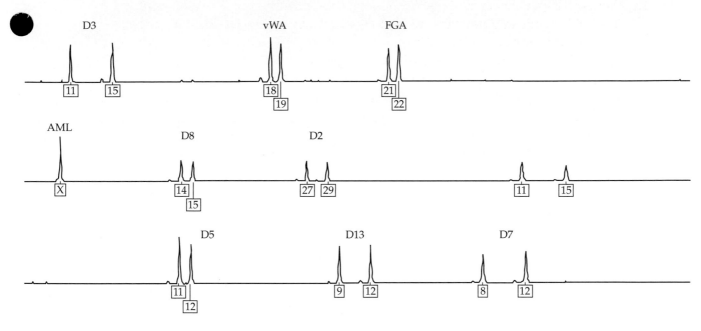

Group 1

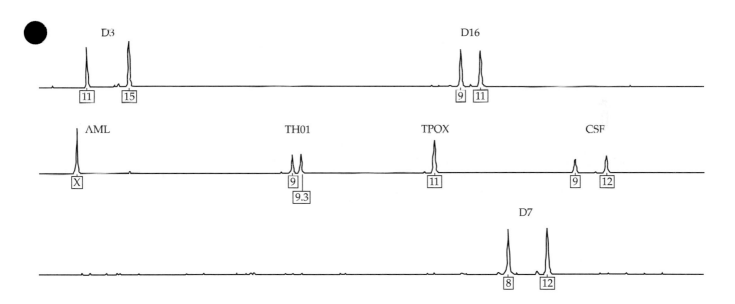

Group 2

FIGURE 18-14. Dog Owner's Blood

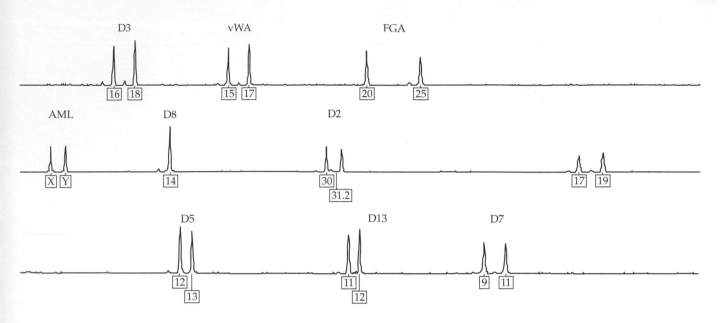

Group 1

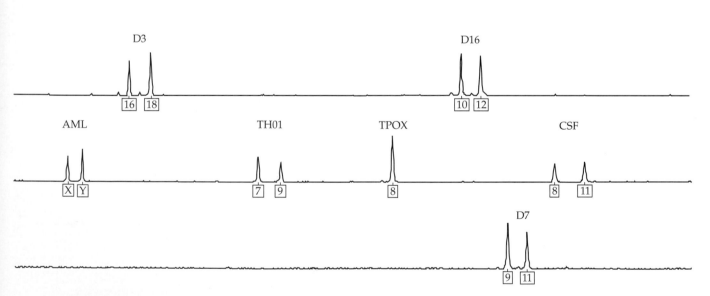

Group 2

FIGURE 18-15. Boy's Blood

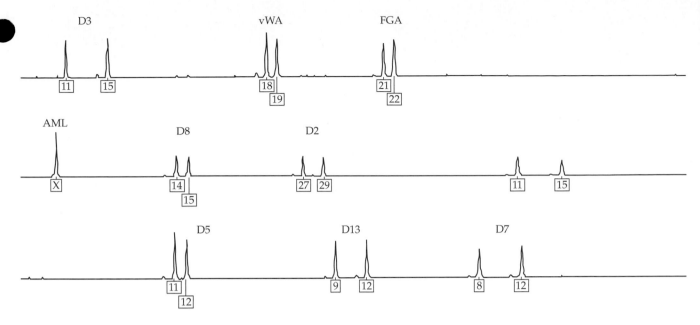

Group 1

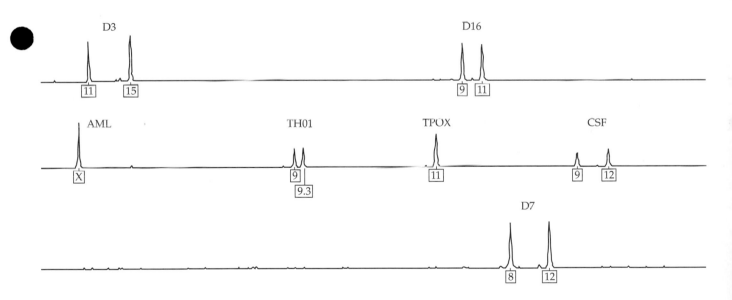

Group 2

FIGURE 18-16. Blood from Dog's Collar

Using Biotechnology to Assess Ecosystem Damage

Objectives

After completing this exercise, you should be able to:

- explain the effects of environmental factors on bioluminescence and give examples
- correctly prepare and use serial dilutions
- demonstrate awareness of factors that can affect the accuracy of experimental results
- explain the utility of serial dilutions to analyze the effects of chemicals on ecosystems
- use experimental results to rank chemicals in terms of their potential to cause harm to human health or the environment
- draw graphs that present data clearly and accurately
- interpret data in tables, charts, and graphs
- apply your knowledge of chemical toxicity to real-life situations.

CONTENT FOCUS

Many types of harmful compounds are released into the environment by industry, agriculture, and even from our own households. It is important not only to detect harmful materials but also to predict their effects on the environment. This is difficult because pollutants are often released in **small amounts, which rapidly disperse** into the soil or water. It is also important to identify **which types** of chemicals are present in the environment, and whether these chemicals are **harmful** to living organisms.

To address this problem, Dr. Kenneth Thomulka, a scientist from Philadelphia, has developed an inexpensive, convenient field-testing method to discover whether toxic chemicals are present in the environment.

The method uses harmless marine bacteria as **biological indicators** for the presence of toxic chemicals.

ACTIVITY 1   EXPLORING BIOLUMINESCENCE

Preparation

The ability of an organism to produce light is called **bioluminescence**. Bioluminescence is exhibited in a variety of organisms, from bacteria to fireflies in your backyard. It is the only light source for marine organisms living deep in the ocean. Like fireflies, some deep-sea fishes use their lights like signals to find mates. Others, like the deep-sea anglerfish, wave glowing lures to attract smaller fish as prey.

Many luminescent organisms, including bacteria, have the enzyme **luciferase**. Luciferase and **oxygen** are needed for the complex chemical reactions used to produce light.

Note:

Read these instructions COMPLETELY BEFORE the lights go out!

---

1. Work in groups. Get the following supplies: **a test tube rack and one 15-ml screw-capped test tube containing a sample of the bacterial culture**.

2. Place the test tube in the rack and set it on your laboratory table. **Unscrew and remove the lid**. Leave the test tube **undisturbed** for 15 minutes.

Caution:

At times during this laboratory period, the room will be completely dark. Make sure the aisles and working areas around your table are clear.

---

3. To demonstrate light production, the class will observe a flask containing **bioluminescent marine bacteria**. With the **lights out**, observe what happens when the instructor swirls the flask containing the bacteria. **Record** your observations.

4. **Continue to observe** the flask for several minutes as it sits undisturbed on the laboratory counter. **Record** your observations.

5. **Without moving or disturbing the test tube rack**, check the tube of bacteria on your laboratory table. Observe the tube **very carefully**.

   Where is the light level **most** intense? _____

   Where is the light level **least** intense? _____

   Using the information presented in the **introduction** to this exercise, which mentions **two** substances needed for bioluminescence to occur, **explain** your observations.

6. **Rank the level of light produced** by the bacteria on a scale of **0 (no light) to 4 (most light)**.

   Top of the tube _____

   Middle of the tube _____

   Bottom of the tube _____

7. **Shake** the test tube **gently** and observe the results. **Describe** what you see.

8. After shaking the test tube, **rank the level of light produced** by the bacteria on a scale of **0 (no light) to 4 (most light)**. _____

   Why did the change in light level occur?

Bacterial test kits are commercially available for detecting pollutants in aquatic ecosystems, such as lakes, streams, and oceans. Manufacturers of household cleaners, shampoos, and cosmetics have started using bacterial testing to replace the more traditional, but highly controversial, tests done with rabbits, rats, and mice.

In the following experiments, you will use bioluminescent bacteria to evaluate the toxicity of common household products. You will try to determine whether any of these household products can be damaging to the environment if they are **not used and disposed of properly**.

Any environmental condition that is harmful to the bacteria's health will cause a decrease in light production.

The point of this experiment is NOT to see how well you can kill bacteria. These bacteria are not harmful. They are valuable and necessary for the environment. The point is to find out how you can PREVENT household chemicals from harming organisms in the environment, including bacteria.

ACTIVITY 2   GETTING STARTED

1. Work in groups. Get the following supplies: **a small flashlight with red filter, one small beaker, one graduated 10-ml pipette with manual dispenser, and one graduated 1-ml pipette with manual dispenser.**

   You will also need: **six large test tubes in a rack, six small test tubes in a rack, scissors, masking tape, a dispenser bottle filled with 3% saline solution, a tray for used glassware, and a container for waste fluids.**

2. Fill the small beaker half full with **tap water**. Attach the manual dispenser to the end of **one graduated 10-ml pipette**.

   Practice filling and dispensing fluid from the pipette following the directions given by your instructor.

3. Now that you are an expert pipette user, you are ready to set up your experiment.

   With masking tape, label the **large** test tubes **D1** through **D6**.

   Make sure you place the tape labels as **close as possible to the top of each tube**.

4. The following household products are available for testing:

    Drain cleaner   Automobile antifreeze

    Mouthwash   Dishwashing liquid

    Household cleaner   Herbicide (weed killer)

    Toilet-bowl cleaner   Pet shampoo

Caution:

Some of these products may be harmful to your skin or eyes. Be careful not to spill any on your hands as you measure. If a spill occurs, DO NOT touch your face or eyes. Wash your hands thoroughly before proceeding.

---

5. Your instructor will assign a household product to your group.

    Obtain a small container filled with your assigned household product.

    Enter the name of the chemical your group will be using: _____

6. You will **dilute** the chemical with various amounts of saline to test the reaction of the bacteria to **different strengths** of this product.

    The dilutions will produce a series of proportionally weaker solutions, commonly referred to as **serial dilutions**. In this manner, you can determine what dilution (concentration) of this chemical is harmful to the bacteria.

    In this experiment, we will compare chemicals to see which are more harmful than others and we will base our assessment of toxicity on the level of dilution necessary to dispose of the chemical safely.

    Rather than just diluting randomly to make this assessment, it is more useful to dilute the chemical in **measured steps (making a serial dilution)**. By exposing bacteria to different concentrations (dilution levels) of the various household chemicals, you can determine the exact level of dilution that will make the chemical harmless.

    Since these are marine bacteria, you will make your dilutions with **3% saline** (a mild salt solution), which simulates the water in the bacteria's natural environment.

7. The dilution levels you will use are listed in **Table 19-1**.

TABLE 19-1 DILUTION SEQUENCE		
TEST TUBE NUMBER	DILUTION	PERCENT CONCENTRATION OF CHEMICAL
D1	Full strength	100
D2	1:10	10
D3	1:100	1
D4	1:1,000	0.1
D5	1:10,000	0.01
D6	No chemical added	0

ACTIVITY 3   MAKING ACCURATE SERIAL DILUTIONS

> **Hint:**
>
> To obtain valid results, you must be very careful to make accurate measurements.

1. Using the **10-ml graduated pipette**, transfer **9 ml** of your assigned household chemical into the first **large** test tube (marked **D1**). This will be the **full-strength** tube.

> **Note:**
>
> Place all used pipettes and glassware into the container provided.

2. Using a **1-ml graduated pipette**, transfer **1 ml** of household chemical **from the container with the original sample** into the next large test tube (marked **D2**).

3. To the **same** tube (**D2**), add **9 ml of saline** solution from the stock bottle. **To dispense the saline,**

    a. hold the test tube under the spout
    b. **slowly** raise the pump handle as far as it will go
    c. **gently** push the handle down to dispense

    Without spilling the contents, **shake the test tube gently** to mix. This will be the **1:10 dilution**.

4. **From test tube D2, remove 1 ml** of solution and transfer it to test tube **D3**.

    Following the instructions above, add **9 ml of saline** to tube **D3**.

    Shake gently to mix. This will be the **1:100 dilution**.

5. **From test tube D3, remove 1 ml** of solution and transfer it to test tube **D4**.

    Following the instructions above, add **9 ml of saline** to tube **D4**.

    Shake gently to mix. This will be the **1:1,000 dilution**.

6. **From test tube D4, remove 1 ml** of solution and transfer it to test tube **D5**.

    Following the instructions above, add **9 ml of saline** to tube **D5**.

    Shake gently to mix. This will be the **1:10,000 dilution**.

    Remove 1 ml of solution from test tube D5 and dispose of it in the **waste container**.

7. To test tube **D6,** add **only 9 ml of saline**. No household chemical will be added to this tube.

✓ Comprehension Check

1. Which test tube contains the **weakest** sample of household chemical? _____

2. Which test tube contains the **strongest** sample? _____

3. Which test tube is the **control**? _____

4. Why is a control tube needed in this experiment?

5. Why was 1 ml of liquid **removed from test tube D5**?

6. **In your own words**, explain the **purpose of making a serial dilution** for this experiment.

Check your answers with your instructor before you continue.

ACTIVITY 4   PREPARING BACTERIAL SAMPLES

1. Get the following supplies: one **clean 1-ml pipette**.

    Using the tube of bacterial culture you looked at during Activity 1, transfer **1 ml of bacteria** to each of the **six small test tubes**.

2. To **each** of the six small test tubes, add **9 ml of saline** solution and set them aside.

Wait!

All laboratory groups must start the next part of the experiment together! While you are waiting, read Activity 5 instructions COMPLETELY.

ACTIVITY 5   TESTING THE TOXICITY OF YOUR CHEMICAL

1. **Record the time** when you begin this procedure. _____

2. **When instructed, carefully empty one small test tube of bacteria into each of the large dilution tubes.**

    Without spilling, **shake each tube gently** to mix the contents.

    Place the empty "bacteria" tubes into the **waste container**.

    It is important to remove these empty tubes from your workspace because the light pollution from the bacteria remaining in the empty tubes will interfere with your observations of the light levels in your experimental tubes.

3. The bacteria/chemical combination must incubate for **15 minutes.**

    Record the time when the incubation period will be completed. _____

4. When the incubation period is completed, observe the bacteria and evaluate the level of bioluminescence present.

Caution:
The room will be COMPLETELY DARK for this activity. Make sure your test tube rack, paper, pen, and flashlight are conveniently positioned before the lights go out!
It will take several minutes for your eyes to become adapted to the dark. To avoid accidents during this period, don't move around.
The flashlight will be only turned on briefly to record your results!

5. **Rank the light** produced by your bacteria according to the **0 through 4 scale** you used in Activity 1 and record your results in **Table 19-2**.

TABLE 19-2

LIGHT INTENSITY OF BACTERIA EXPOSED TO SERIAL DILUTIONS OF ONE HOUSEHOLD CHEMICAL (0–4 SCALE)

TEST TUBE NUMBER	DILUTION	RANKING (0–4)
D1	Full strength	
D2	1:10	
D3	1:100	
D4	1:1,000	
D5	1:10,000	
D6	No chemical added	

6. **Record** the results for your group on the **master chart** at the front of the room.

From the master chart, copy the results for the entire class into **Table 19-3**.

T A B L E 1 9 - 3								
LIGHT INTENSITY OF BACTERIA EXPOSED TO SERIAL DILUTIONS OF ALL EIGHT HOUSEHOLD CHEMICALS (0–4 SCALE)								
TUBE #	DRAIN CLEANER	MOUTH-WASH	HOUSE-HOLD CLEANER	ANTI-FREEZE	DISH-WASHING LIQUID	WEED KILLER	TOILET-BOWL CLEANER	PET SHAMPOO
D1 (full strength)								
D2 (1:10)								
D3 (1:100)								
D4 (1:1,000)								
D5 (1:10,000)								
D6 (control)								

7. Using the data from **Table 19-3**, make **eight bar graphs** that show the results of each experiment in **Figure 9-1.**

Label each graph with the name of the household chemical that was tested.

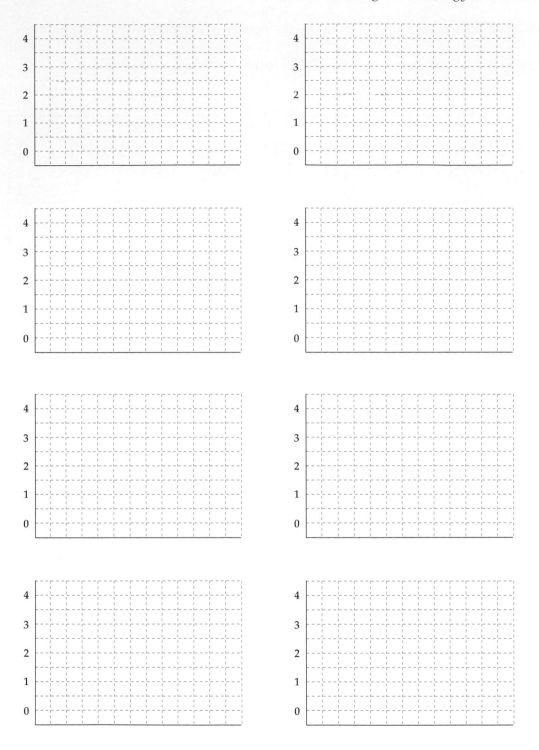

FIGURE 19-1.   Comparisons of Test Results for Eight Household Chemicals

Check your graphs with your instructor before you continue.

✓ Comprehension Check

1. Which household chemical was the **most toxic**? _____

   Why did you conclude that this chemical was the most toxic? In your explanation, **include facts** collected during the bacteria experiments.

2. What part(s) of the food chain might be affected if this chemical was disposed of improperly in the environment? **Explain your answer.**

3. Imagine this situation. A type of drain cleaner is found to be harmless to bacteria if it is diluted in a 1:10,000 ratio. A 1:10,000 dilution is the same as **1 milliliter of chemical diluted by 10 liters of water**. The drain cleaner container holds 200 ml of chemical. If the directions tell you to pour the entire contents of the container down the drain, how **many liters of water** would be needed to dilute this chemical to safe levels?

   _____ liters   **Show your work.**

4. In an experiment testing the toxicity of a car wash chemical, bacteria exposed to the chemical at a **1:1,000 dilution** showed a light level of **two**. Bacteria exposed to a **1:10,000 dilution** of the same chemical showed a light level of **three**.

    In order to dispose of this chemical safely, is a 1:10,000 dilution sufficient? **Explain your answer.**

5. What would you conclude if **two of the tested products** showed **NO biolumi-nescence** at a **1:1,000 dilution**? **Explain your answer.**

6. What would you conclude from your experiment if **two** of the tested household products showed the **same light intensity at a 1:10,000 dilution**? **Explain your answer.**

7. *Challenge Question!* Can the bioluminescence of bacteria exposed to a **1:10,000 dilution** of a chemical ever equal the amount of bioluminescence in the **control tube**? **Explain your answer.**

Check your answers with your instructor before you continue.

Self Test

Figure 19-2 contains the results of a bioluminescent bacteria assay performed with two lawn-care products that frequently find their way into sewers, streams, lakes, and ponds.

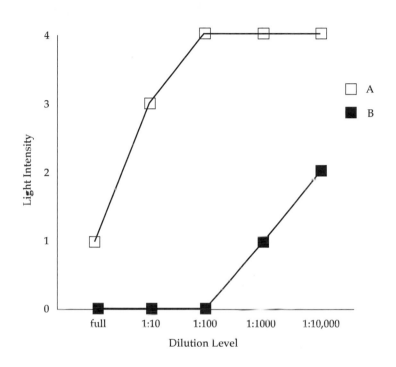

FIGURE 19-2. Bioluminescent Bacterial Assay of Two Lawn-Care Products

1.   At which dilution level was **Product A** most toxic? _____

What is the **lowest** dilution level at which **Product A** was harmless? _____

2.   At which dilution level on the graph is **Product B** harmless? _____ **Explain your answer.**

3.   If these two lawn-care products do the same job, which product is less harmful to the environment? **Explain your answer.**

E X E R C I S E

20

Population Ecology

Objectives

After completing this exercise, you should be able to:

- estimate the size of a population using various methods of sampling
- use ecological quadrats as a tool for predicting population size
- calculate the percent error of your estimate
- draw conclusions that are supported by experimental data
- apply your knowledge of the scientific method to real-life situations.

CONTENT FOCUS

In studying a population in an ecosystem, it is usually essential to know the size of the population. Ecologists call the total count of all the individuals in a population a **census**.

Scientists cannot possibly count every organism in a population. One way to estimate the size of a population is to collect data by taking random samples. In this activity, you will look at how data obtained by random sampling compare with data obtained by an actual count. Sampling is used to track population growth in an ecosystem. It is one of many methods used by scientists to collect data when studying ecosystems.

ACTIVITY 1

INTRODUCTION TO SAMPLING— AN ECOLOGICAL TOOL

Preparation

Estimates of population size are basic to understanding the interactions of populations in an ecosystem. It is seldom practical, however, to count every individual in a population. Even if possible, direct counts may be too time-consuming or expensive. Sampling is a way to make a quick estimate of population size. To give you an idea of how an ecologist would sample a wild population, we have taken a photograph of a flock of birds.

1. Work in groups. Consider the population in **Figure 20-1**. It would be hard to count the exact number of birds in this flock, especially if the flock was moving, so a quick estimate is our best option.

2. Look at the population in **Figure 20-1**. If you stretch your imagination, you will see that the area covered by birds can be roughly divided into six squares.

   DON'T draw grid lines on the photograph—you are simulating what you would be able to do if you were actually outside counting the birds!

3. To make a population estimate, count the number of birds in one square and multiply the total by six.

   _____ × 6 = _____ (total population)

4. How close did you come to the actual population? Count every bird in the photograph to find out.

   Exact count of total population = _____
   Check one of the following to rate your estimating accuracy:
      Within **5 birds** of the actual total   Accuracy **100%**
      Within **25 birds** of the actual total   Accuracy **90%**
      Within **50 birds** of the actual total   Accuracy **80%**
      Within **70 birds** of the actual total   Accuracy **70%**
      Within **100 birds** of the actual total   Accuracy **60%**

✔ Comprehension Check

1. How does each of the following factors affect the accuracy of population estimates? Explain each answer.

   a. Practicing making estimates:

   b. Making sure your "counting" square contains a representative number of animals:

   c. Averaging the results of two squares or three "counting" squares:

2. What are the advantages of estimating population size using the "imaginary counting square" method? What are the disadvantages?

FIGURE 20-1.   Birds on the Wing

ACTIVITY 2

SAMPLING QUADRATS TO ESTIMATE POPULATION

Preparation

You are a student in an ecology class that has taken a field trip to Yellowstone National Park. Each grid square in **Figures 20-3 and 20-4** represents a study area within the park. Each study area is divided into 100 **quadrats** for vegetation sampling (see **Figure 20-2** for an example of how a quadrat is established).

FIGURE 20-2.   Example of a Quadrat

You will be engaged in a habitat assessment of the meadow community in the study area. Habitat assessment is a common tool used to obtain an overall evaluation of a habitat at relatively low cost. With baseline data obtained by habitat assessment, we can predict the environmental impact of activities proposed for this location.

We are beginning with an assessment of plant populations. Since plants are fixed in position, we can establish measured grids to facilitate estimates of population size and distribution.

Building on the random sampling method you practiced in Activity 1, you will try to make an accurate estimate of the population of gold flowers (*Hymenoxys acaulis*).

1. Work in groups. Get **one set of 10 alphabetic chips and one set of 10 numeric chips**. Each set of chips will be in a small container.

2. To begin sampling, randomly remove **one chip from each container**. Record the number and letter you drew in **Table 20-1** in the column titled **Quadrats Sampled in Figure 20-3**. For example: **F8**.

   Return the chips to the container before you draw another chip combination.

   Continue drawing chips from the two containers and recording the numbers until you have identified **10 quadrats to sample**.

3. To sample the first quadrat on your list, locate the quadrat in your grid square. Count all the plants in that square (represented by dots on the grid) and record the total next to the appropriate quadrat number in **Table 20-1**.

> ### Note:
>
> What should you do in case a dot is on the line between two quadrats? If at least half of the dot is in the quadrat you are sampling, count it as part of your total for that quadrat.
>
> **If you count a dot that is on a line,** mark through that dot **so that you don't mistakenly count it again as part of the population in an adjoining quadrat.**

Continue with this procedure until you have sampled and recorded the data for all ten quadrats.

4. **Repeat Steps 2 and 3** above to sample the population in **Figure 20-4**.

   Return to **Table 20-1** and record the quadrat numbers you will sample in the column titled **Quadrats to Sample in Figure 20-4**. Record your population results in the **Plants Counted** column for **Figure 20-4**.

5. Record the total number of plants counted in all 10 quadrats and record the data in **Table 20-1**.

   To estimate the total population size in your grid square, **multiply the total number of plants counted by 10**. Record your population estimate in **Table 20-1**.

6. Return to the grid square and count every dot on the grid (an actual count of the population present in your grid square).

   Record the actual population count in **Table 20-1**.

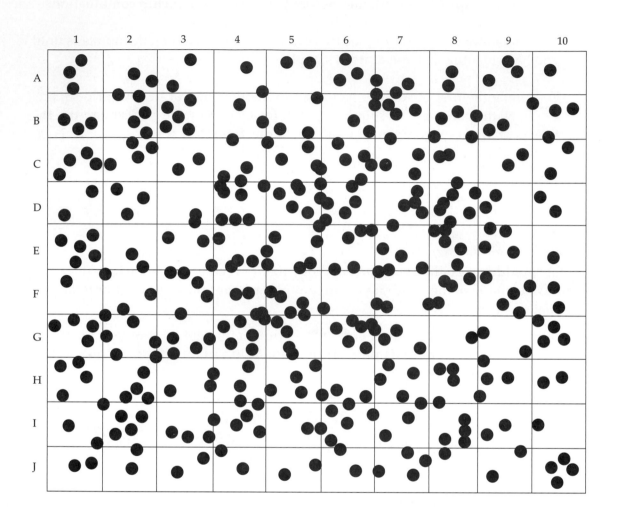

FIGURE 20-3. Study Area 1

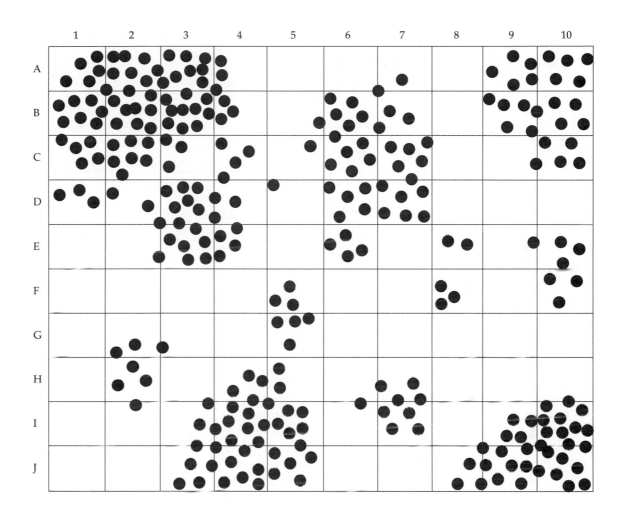

FIGURE 20-4. Study Area 2

7. Use the following formula to calculate the percent of error between your population estimate and the actual population.

$$\text{percent error} = \frac{\text{actual population} - \text{estimated population}}{\text{actual population}} \times 100$$

Record your percent error in **Table 20-1**.

TABLE 20-1			
VEGETATION SAMPLING DATA			
QUADRATS SAMPLED IN FIGURE 20-3	PLANTS COUNTED IN FIGURE 20-3	QUADRATS SAMPLED IN FIGURE 20-4	PLANTS COUNTED IN FIGURE 20-4
	Total Plants Counted:		**Total Plants Counted:**
	Population Estimate from Sample:		**Population Estimate from Sample:**
	Actual Population Count:		**Actual Population Count:**
	Percent Error of Estimate:		**Percent Error of Estimate:**

✓ Comprehension Check

1. Was there a difference in the percent error of your population estimate between **Figures 20-3 and 20-4**?

    If so, how might the following factors have contributed to the difference?

    a. Distribution of plants in the quadrats (evenly distributed vs. patches of plants):

    b. Number of quadrats sampled:

2. Why did we use the "chip" method to select the quadrats to be sampled instead of simply selecting 10 "likely looking" quadrats?

3. List and explain several environmental factors that might contribute to the patchy distribution of the plants in **Figure 20-4**.

4. List **three situations** in which **random sampling of human populations** is used to collect information.

✓ Challenge Question!

5. You notice, while passing outside your local elementary school, that almost all the parents waiting to pick up their children were driving SUVs. You wonder if the same distribution of vehicles would be observed at other elementary schools in the United States.

   Design an experiment that uses sampling techniques to answer this question. **Explain** your method and the reasoning behind your experimental design in detail.

Check your answers with your instructor before you continue.

ACTIVITY 3

THE MARK AND RECAPTURE TECHNIQUE
FOR ESTIMATING POPULATION SIZE

Preparation

If you are trying to sample a population of animals that can move from place to place, the quadrat system won't be very effective. To sample this type of population, individuals are captured in a study area, marked for identification, and released. Later, individuals will again be captured in the same study area. Some of these newly captured individuals will be marked, others will not. On the basis of the ratio between marked and unmarked individuals captured, it is possible to make a mathematical estimate of the population size.

The following activity is a simulation of the mark and recapture technique demonstrated by sampling a captive population of brine shrimp.

1. Work in groups. Get **a container of brine shrimp in clear saline solution (labeled "brine shrimp population"). You will also need a small beaker of brine shrimp that have been colored with methylene blue dye, several disposable pipettes, and two empty beakers**.

   Label the first beaker #1 and the second beaker #2.

2. **Capture Phase**

   Capture **30 brine shrimp**, using the following method.

   To "capture" the brine shrimp, fill a pipette in a **random location** in the "population" container.

   When you capture a sample of brine shrimp, you will place the captured individuals in **Beaker #1**.

3. **Marking Phase**

   Since it would be almost impossible to mark an animal as small as a brine shrimp during one lab period, we will use an alternate method of releasing marked individuals into the population.

   Replace each brine shrimp captured and removed from the population with a blue brine shrimp (you will be releasing a total of 30 blue individuals into the original brine shrimp population).

   Wait for five minutes to allow the marked individuals to become randomly dispersed in the population.

4. **Recapture Phase**

Your recapture sampling of the population **MUST BE RANDOM. WITHOUT LOOKING**, insert the pipette into the population container and draw up a sample.

Count the number of marked and unmarked brine shrimp in the pipette.

Note:

If counting the shrimp in the pipette proves difficult, release the contents of the pipette into *Beaker #2 and then count the shrimp.* **By placing the beaker on a white sheet of paper, the brine shrimp will be easier to observe and count.**

5. **Record the number of marked and unmarked individuals captured in the first pipette in Table 20-2.**

6. After you have recorded your shrimp count, transfer the counted shrimp to **Beaker #1**.

   DO NOT REPLACE THE SAMPLED SHRIMP IN THE "POPULATION" CONTAINER.

   Continue taking random recapture samples and recording your data until you have recaptured **10 marked individuals**.

TABLE 20-2
RESULTS OF MARK AND RECAPTURE EXPERIMENT

PIPETTE SAMPLE TAKEN	NUMBER OF UNMARKED SHRIMP CAPTURED	NUMBER OF MARKED SHRIMP RECAPTURED
1.		
	Total:	Total:

7. Use the following formula to calculate your population estimate:

    N = estimate **of total population size**

    C = total number **of shrimp captured (marked and unmarked)**

    M = total number **of shrimp marked**

    R = number **of marked shrimp recaptured**

    $$N = \frac{(C)(M)}{R}$$

    On the basis of your calculations, what is your shrimp population estimate? _____

    What is the probability that your mark and recapture technique gave you an accurate population estimate?

    The degree of accuracy can be calculated based on a principle known as the **95% confidence interval**. The confidence interval calculations produce a range of numbers (high and low) between which the actual population figure should lie (with 95% confidence that this is correct).

✔ Comprehension Check

1. **(Circle one answer.)** If the marked individuals were more visible to predators than the unmarked individuals my population estimate would be **too high/too low. Explain your answer.**

2. **(Circle one answer.)** If marked animals migrated out of my study area, my population estimate would be **too high/too low. Explain your answer.**

3. Fishery managers have used population estimates to determine the maximum sustainable yield (the degree of harvesting that does not harm the populations' long-term survival). What would be the benefits and the shortcomings of using the **mark and recapture technique** to estimate the number of tuna that commercial fishermen are allowed to capture each year?

Self Test

1. At a forested site, a lumber company sampled the number of 12-inch diameter pine trees in preparation for logging. Later, in a direct count, it was discovered that they had seriously underestimated the number of suitable trees. What factors may have contributed to the lack of estimating accuracy? **Explain your answer.**

2. Several state and federal agencies support programs to tag and release commercially important fishes. For example, sport fishermen tag and release the large fish they catch. If someone catches a tagged fish at a later date, they are asked to report the geographic location, the size of the fish, time of the year, and other information. How can this information be used to benefit the tagged species?

Self-Test Answers

Exercise 1—Introduction to the Scientific Method

1.   B - state hypothesis

2.   C - state results (facts only)

3.   A - test hypothesis (by experiment or observation)

4.   A - test hypothesis (by experiment or observation)

5.   B - state hypothesis

6.   B - state hypothesis

7.   C - state results (facts only)

8.   a.   The points are too closely spaced together (they should be spread out as much as possible along the Y-axis) and crowded into the corner (they should be evenly spaced along the X-axis)

         b.   No numbers or information about scale on the X-axis

         c.   No graph title

9.   a.   Missing titles from the X- and Y-axes

         b.   Numbers on the Y-axis do not have equal intervals; the scale changes from intervals of 50 (50–100) to intervals of 500 (100–500)

         c.   No numbers or information about scale on the X-axis

         d.   Plotted points extend above the number scale on the Y-axis

10. a. Missing titles from the X- or Y-axes

     b. No numbers or information about scale on the X-axis

11. Decreased

12. About 13%

13. About 88%

14. 1980

Exercise 2—Interdependency Among Organisms

1. Producer

2. Examples: algae, phytoplankton, pine tree, rose bush, corn, rice, beans, photosynthetic bacteria (any organism that contains chlorophyll is acceptable)

3. Primary consumer

4. Secondary consumer

5. Weeds ⟶ crickets ⟶ rats ⟶ cats

    producer   primary   secondary   tertiary
                consumer   consumer   consumer

6. If one of the cats died in the vacant lot, it would be consumed by decomposer organisms. In this way, nutrients contained in the cat's body would be recycled and the decomposers would obtain energy.

7. Some plants store starch in their roots. Examples include potatoes, carrots, radishes, turnips, and beets.

8. On the basis of the methods used in this experiment, the conclusions are not valid. When the refrigerator door is closed, the light is out. Since two variables are involved (temperature and light), it is impossible to tell which caused the observed effects.

Exercise 3—Windows to a Microscopic World

1.

Ocular lens (eyepiece)	Allows you to view the specimen; magnifies image by 10 times (10×); contains pointer
Stage	Holds the specimen; has a mechanism to move the slide around
Condenser lens	Focuses light on the specimen
Nosepiece	Revolves to change objective lenses
Scanning lens (4×)	Objective lens used to first locate a specimen
Iris diaphragm	Regulates the amount of light that passes through the specimen
Fine focus knob	Moves objective lenses in small increments for small adjustments in image clarity
Scanning lens (4×)	Objective lens with the lowest magnifying power
High power (40×)	Objective lens with the highest magnifying power
Coarse focus knob	Moves objective lenses rapidly in large increments for initial adjustment in image clarity

2.   Ocular lens = 15   Objective lens = 20

   Total magnification: $15 \times 20 = 300$

3.   In the compound microscope, the light source is beneath the specimen. Therefore, only thin specimens can be viewed. The dissecting microscope can provide light from many different directions so that large, thick objects can be viewed.

   In the compound microscope, the image is viewed upside down and backward. This is not true for a dissecting microscope.

   The dissecting microscope has a zoom lens that can gradually increase magnification. The lenses of the compound microscope have fixed magnification levels.

   The compound microscope has much greater magnification ability than a dissecting microscope.

4.   Compound—cells from stomach lining

Dissecting—sea shell

Dissecting—cockroach

Compound—mildew

5.   The daphnia moved away from you and to your left (because images in the compound microscope are inverted and reversed).

6.   a.   Go back to 10× and center the specimen in the middle of the field of view or check to make sure the 40× objective is clicked into position.

b.   Adjust the iris diaphragm to let in more light.

c.   Check to make sure the objective lens is clicked into position, or check to see if the objective and ocular lenses are clean.

d.   Use lens paper to clean the slide, the objective lenses, and the ocular lens.

e.   These are air bubbles. Lift the cover slip and lower it gently at a 45° angle to expel the air.

Exercise 4—Functions and Properties of Cells

1.   Cell wall, chloroplast, sap vacuole

2.   Arrow should show water moving from the bag into the beaker.

3.   **Dialysis** is the separation of different sized molecules through a selectively permeable membrane. As diffusion takes place across the membrane, the composition of the blood changes. Waste products such as phosphate ions, urea, and potassium ions cross the membrane into the dialysis fluid. Blood cells, proteins, and other large molecules cannot cross the membrane and are retained in the blood.

4.   Distilled water has no dissolved solutes. Osmosis refers to the movement of water molecules from an area of high concentration to an area of lower concentration. In the described situation, there is a **lower concentration** of water molecules inside the cell (in the cytoplasm) and a **higher concentration** of water outside the cell (the distilled water). Through the process of **osmosis**, the distilled water will cross the **cell membrane** and enter the blood cells, causing them to swell and rupture the cell membrane.

5.   The sugar cube will dissolve faster in the hot tea. During **diffusion**, molecules move from an area of high concentration to an area of lower concentration. Since the water is hot, molecular movement increases. **Collisions** between molecules become more frequent. As molecules bump into each other, they **diffuse** outward, spreading from the area of **high sugar concentration** (the sugar cube) to an area of **lower sugar concentration** (the cup).

6.   The smoke **molecules** are moving by **diffusion** over the restaurant barrier from an area of **high smoke concentration** (the smoking section) to an area of **lower smoke concentration** (the nonsmoking area).

Exercise 5—Investigating Cellular Respiration

1.   Bromothymol blue turns yellow when CO_2 is present. Since the solution did not change color, we can conclude that cell respiration did not take place in the container (no CO_2 was produced). In this situation, we cannot even be sure the green spots were plants.

2.   a.   The cloudy result indicates that the gas obtained from the body cavity contained CO_2.

      b. CO_2 is a byproduct of cellular respiration. Cell respiration occurs only in living organisms. Since the animal is dead, the CO_2 produced must be coming from the cellular respiration of other organisms (decomposers present on the carcass).

3.   A—aerobic respiration

      Aerobic respiration produces about 36 ATPs per glucose molecule, whereas anaerobic forms of cellular respiration produce only 2 ATPs per glucose molecule.

4.   Bromothymol blue returned from yellow to its original blue color because CO_2 was no longer present. Since plants absorb CO_2 during photosynthesis, and the plant was exposed to sunlight during the several hours in question, this is evidence that photosynthesis occurred.

5.   B—only under anaerobic conditions

      In the absence of sufficient oxygen, yeast will shift their method of cellular respiration to an anaerobic method—alcohol fermentation. In this way, they can continue to obtain ATP to support cellular activities.

6.   C—under both aerobic and anaerobic conditions

Breakdown of glucose for cell respiration begins with glycolysis, an anaerobic process.

Cellular respiration will continue as shown in **Figure 3**, but the pathway used is determined by the presence or absence of oxygen.

7.   B—only under anaerobic conditions

During strenuous exercise, when sufficient oxygen is not available to muscle cells, cellular respiration shifts to an anaerobic method—lactic acid fermentation. In this way, muscle cells can continue to obtain ATP to support cellular activities.

8.   A—aerobic conditions only

Under ideal conditions, cells can maximize their ATP production.

9.   D—Even though anaerobic respiration produces small amounts of ATP, the amount is not sufficient to meet the energy needs of an animal for long periods of time.

Exercise 6—Nutrient Analysis of Foods

1.   Your morning orange juice doesn't contain protein.

2.   Since potatoes contain large amounts of starch, the iodine would probably turn black (a positive starch test).

Since animal tissues do not contain starch, the iodine test should be negative (no color change).

3.

Peanut	a seed; contains the embryo and food reserved
Tuna	skeletal muscle tissue from a fish; produces voluntary movements
Bean	a seed; contains the embryo and food reserved
Milk	liquid high in nutrients; produced by mammary glands of mammals to feed their offspring
Hamburger meat	skeletal muscle tissue from a cow; produces voluntary movements
Lettuce	leaf of lettuce plant; site of photosynthesis

Exercise 7—Factors That Affect Enzyme Activity

1.   We have seen that unusually high temperatures can affect the three-dimensional folding of a protein. In enzymes, if the structure of the active site is altered, the enzymes' ability to function may be destroyed. High fevers may affect enzymes throughout the body.

2.   Enzymes are sensitive to changes in environmental conditions and many are quite specific as to the pH range in which they can function.

3.   3.5

4.   At pH 3.7, it took 14 minutes before noticeable enzyme activity was observed. In comparison, enzyme activity was observed after only two minutes at pH 3.5.

5.   The enzyme is specific because it's optimal function was centered around pH 3.5. Enzyme activity at pH values even slightly above or below 3.5 were greatly decreased.

6.   The production of melanin (and thus, the fur color in Siamese cats and Himalayan rabbits) is related to body temperature. This trait does not apply to colors of skin, fur, and feathers in all animals, just these specific breeds.

6.   Blood circulation and body temperature both decrease in the extremities. Since body temperature is lower in the paws, ear tips, and nose, the enzyme for melanin production is active. Core temperatures at the center of the body are too high for melanin synthesis to be completed.

Exercise 8—Functions of Tissues and Organs I

1.   I - sebaceous gland

2.   J - dermis

3.   A - keratin

4.   D - epithelial cells

5.   C - erector muscle

6.   H - subcutaneous layer

7.   B - melanin

8.   E - epidermis

9.   No. Lotions applied to the outer skin surface do not penetrate to the lower layers of the epidermis where new skin cells are produced.

10.   The nicotine medication must penetrate through the epidermis and enter blood vessels in the dermis.

11. On the sole of the foot, the epidermal layer is thickened and calluses may be present. Penetration of the patch medication through to the dermis might prove difficult.

12. In second-degree burns, the epidermis has been destroyed, exposing the sensitive nerve endings in the dermis. Pain will be severe. In third-degree burns, nerve endings are completely destroyed. Consequently, the burn patient feels no pain until the nerves begin to regenerate and repair themselves.

Exercise 9—Functions of Tissues and Organs II

1. D - yellow marrow

2. E - red marrow

3. G - compact bone

4. F - smooth muscle

5. C - cardiac muscle

6. B - skeletal muscle

7. B - skeletal muscle

8. A - spongy bone

9. H - shaft

10. G - compact bone

11. Your friend is experiencing muscle fatigue. Actively contracting muscles become weaker as time passes. This can be caused by lack of ATP, insufficient oxygen, depletion of energy reserves in the muscle cells, and accumulation of metabolic wastes.

12. In the intense heat, the protein fibers in the bone matrix were destroyed. Protein fibers make an important contribution to the strength and resiliency of a bone. Without them, the bone becomes brittle and easily crushed.

13. Tendon

14. Ligament

Exercise 10—Introduction to Anatomy: Dissecting the Fetal Pig

1. C - head end of the body

2. A - back

3.   B - tail end of the body

4.   D - belly side

5.   Female pigs have a genital papilla just ventral to the anus. In addition, the uro-genital opening of females is located just below the papilla (as opposed to males, in which the urogenital opening is located just posterior to the umbilical cord).

6.   D—The trachea is the only air passageway to the lungs. Although the esophagus is also located in the throat, it is not involved in the breathing process.

7.   Esophagus

8.   Diaphragm

9.   Under heavy exercise, a person with emphysema will not get enough oxygen for the body's needs. The maximum amount of air moved in and out (**vital capacity**) when breathing deeply is significantly reduced.

10.   When insufficient **oxygen** is transported by **red blood cells**, the **mitochondria** of other body cells cannot function at peak efficiency. The amount of **ATP** produced decreases and, therefore, the amount of **energy** available to power cell activities will also decrease.

Exercise 11—Organs of the Abdominal Cavity

1.   H - mesenteries

2.   C - villi

3.   D - circular muscle

4.   F - peristalsis

5.   E - longitudinal muscle

6.   B - root hairs

7.   Digestive system: stomach, small intestine, large intestine, pancreas, liver, gall bladder

Respiratory system: no organs from this system in the abdominal cavity

Urinary system: kidney, bladder, ureter

Circulatory system: arteries, veins, capillaries

8.   Cardiac sphincter; stomach acids back up and irritate the lining of the esophagus. Since this region of the esophagus is not far from the heart, this discomfort can be mistaken for chest pains.

9.   Stomach acids react with the enamel on the teeth and gradually dissolve the enamel over long periods of time.

10. The malfunctioning organ is probably the bladder. The malfunctioning structure is probably the sphincter associated with the bladder and urethra, which prevents urine from being released accidentally.

11. Design B will be more efficient because more surface area is in contact with the air.

12. Crossword puzzle answers:

Exercise 12—The Circulatory System

1. 1 - tissue capillaries in big toe

    6 - pulmonary artery

    9 - left atrium

    8 - pulmonary vein

    4 - right atrium

    10 - left ventricle

    3 - inferior vena cava

    11 - aorta

    5 - right ventricle

    12 - arterioles

    2 - venules

    7 - lungs

2. Blood would back up into the right atrium.

3. 96

4. O - tissue capillaries entering big toe

    D - pulmonary artery

    O - left atrium

    O - pulmonary vein

    D - right atrium

O - left ventricle

D - inferior vena cava

O - aorta

D - right ventricle

D - tissue capillaries leaving big toe

O - arterioles

D - venules

5. Owing to rheumatic fever, scarred heart valves are not closing completely. The hissing sound is caused by a small amount of blood leaking through the valve under pressure when the ventricles contract.

6. Skin arteries dilate (enlarge) to increase blood flow to the skin.

7. Right ventricle—40   pulmonary artery—40

    Pulmonary vein—100   left atrium—100

8. Blood flow to the muscles increases from about 21% to about 73%.

Diameter of blood vessels supplying muscle tissue also increases.

9. Blood flow to the abdominal organs decreases from about 25% to about 3%.

Diameter of blood vessels supplying abdominal organs also decreases.

10. Blood flow to the skin increases from about 9% to about 12%.

Increased blood flow to the skin is one mechanism by which the body releases excess heat during exercise.

Exercise 13—Introduction To Forensic Biology

1. C - whorl

2. G - divergence

3. F - bifurcation

4. E - loop

5. J - sweat glands

6. H - fingerprint formula

7. I - points of similarity

8. D - tented arch

9. The skin is covered with sweat glands that produce perspiration that can accumulate on the ridges forming the fingerprints. These ridges also accumulate body oils from touching oily surfaces. Together, these leave an invisible impression, which is the fingerprint.

10. Even though their fingerprints might be the same at birth, differing daily activity cause an accumulation of scars and other marks that can be used to tell the fingerprints of the twins apart.

11. The blood type is A–. Agglutination with anti-A serum shows the presence of type "A" proteins, but lack of agglutination with anti-B and anti-Rh serums shows that these proteins are absent.

12. The blood type is AB–. Agglutination occurred with both anti-A and anti-B serums, showing the presence of type "A" and type "B" proteins, but lack of agglutination with anti-Rh serums shows that this protein is absent.

13. The blood type is O+. Agglutination did not occur with either anti-A or anti-B serums, showing the absence of type "A" and type "B" proteins, but agglutination did occur with anti-Rh serum, which shows that the Rh protein is present.

Exercise 14—Mitosis and Asexual Reproduction

1. E - nuclear membrane

2. I - equator

3. C - centromere

4. K - daughter cells

5. A - diploid

6. D - spindle fibers

7. H - cleavage furrow

8. B - sister chromatid

9.   G - cytokinesis

parent
(2n = 4)

interphase
(chromosomes
duplicated)

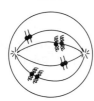
late prophase
(chromosomes attach
to spindle fibers)

metaphase
(chromosomes line
up along equator)

anaphase
(sister chromatids
separate)

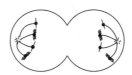
telophase
(cytoplasm
divides)

daughter cells
(2n = 4)

10.   L - chromosome

11.   False

12.   True

13.   True

14.   True

15.   Stages of mitosis for parent cell with four chromosomes:

Prophase I	synapsis and crossing-over
Telophase I	two haploid daughter cells first appear
Metaphase I	homologous chromosomes line up on the equator **in pairs**
Metaphase II	chromosomes line up on the equator (there is only one chromosome from each homologous pair)
Interphase	chromosome replication **before** meiosis begins
Anaphase II	centromere splits and sister chromatids separate
Telophase I and II	division of cytoplasm occurs
Telophase II	four haploid daughter cells first appear
Anaphase I	pairs of homologous chromosomes separate
Anaphase II	sister chromatids separate

16.   The plan will not be effective. The fishermen are actually doubling the number of starfish. Starfish are capable of regeneration, which is an example of asexual cell division. Each half of the starfish can replace lost body tissues and develop into a new, fully functional adult.

17.   A—Skin cells are diploid and have a full set of chromosomes necessary to provide the genetic information to make a new body. The other two options are not viable since sperm cells are haploid and red blood cells have no nucleus.

Exercise 15—Connecting Meiosis and Genetics

1.

2.
The father	Aa
The mother	Aa
The child	aa
Probability of an albino child	25%

Probability of normally pigmented child   75%

Explanation: A child inherits **one** allele from each parent; therefore, each parent must have one little "a."

3.   Mother   Tt

        Father   Tt

        First child   tt

Explanation: A child inherits **one** allele from each parent; therefore, each parent must have one little "t." The parents could not, however, be homozygous recessive, since people with Tay Sachs disease die in early childhood.

4.   Ghandi   Cc

        Sabrina   Cc

        Snowflake   cc

        Probability of heterozygous cub   50%

5.   Ghandi   Cc

        White tiger at zoo   cc

        Probability of another white cub   50%

Exercise 16—Useful Applications of Genetics

1.   The woman   tt

        The man   Tt

        The two sons   Tt

        The daughter   tt

        Grandparents   Tt and T?

2.   Your genotype   Dd

        Your husband   DD

        Your sister   dd

        Probability of Tay Sachs child   0%

3. Fred X^BY^o

   Ginger X^BX^b

   David X^bY^o

   Takiyah $X^BX^?$ but probably X^BX^B

   Kelly X^BX^b

   Kevin X^bY^o

   Takiyah's five sons X^BY^o

   Probability of color-blind son   25%

   Probability of color-blind daughter   0%

4. Ralph X^HY^o

   Ralph's brothers X^hY^o

   Ralph's sister X^hX^h

   Ralph's mother X^HX^h

   Ralph's father X^hY^o

   Since Ralph has normal vision, he must have inherited a big "H" from his mother. Since Ralph has a color-blind sister, however, his mother must be heterozygous (X^HX^h). His father must be color-blind (in order to produce a daughter with the genotype X^hX^h).

5. Pedigree showing albino individuals:

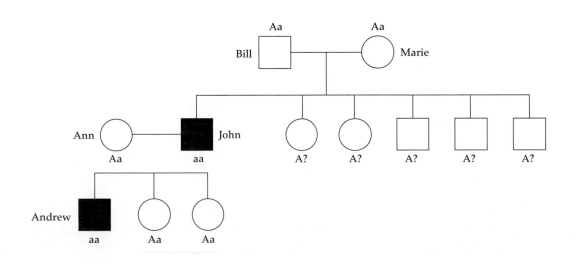

Exercise 17—Introduction to Molecular Genetics
I've got it—the genetic code! Wait a minute. Who alphabetized this?!

Exercise 18—Biotechnology: DNA Analysis

1.   Junk DNA is that portion of the DNA molecule that has no known function and does not code for the synthesis of proteins.

2.   The pattern of repeating junk DNA sequences differs from person to person. This gives each person a unique DNA profile through the process of RFLP analysis.

3.   No child is genetically identical to either of the parents. Through the process of meiosis and fertilization, each child inherits half the genes from each parent.

      By analyzing the child's STR profile and comparing it to both parents, it is possible to determine which bands were not inherited from the mother, and therefore that they must have been inherited from the father.

      In blood sample comparisons, however, the entire STR profile must be a match between the two samples to show that the blood samples both came from the same person.

4.   The deer meat could be tested by STR analysis. If the deer's STR profile was matched to skin or internal organs found inside the park, this would demonstrate conclusively that the confiscated deer was the individual killed within the park boundaries.

5.   The blood found on the dog's collar matches the owner's blood. An analysis of the STR profile of the child's blood does not match the sample from the dog's collar. This evidence substantiates the owner's claim that his dog was not responsible for the attack.

Exercise 19—Using Biotechnology to Assess Ecosystem Damage

1.   Product A was most toxic at full strength because bacteria only glowed at an intensity level of one.

      Product A was harmless at 1:100 because this was the first dilution at which bacteria glowed at an intensity level of four.

2.   There was no dilution level at which Product B was harmless. Even at the lowest dilution (1:10,000), the bacteria only glowed at an intensity level of two.

3.   Product A is least harmful to the environment because it is less toxic at lower dilution levels. It can be safely disposed of by mixing with water in a dilution of 1:100.

Exercise 20—Population Ecology

1.   Accuracy would be affected if the trees had a patchy (rather than even) distribution in the forest. In this situation, your sampled quadrats might not be truly representative of the tree population and your estimate would be too low. One cause of patchy distribution would be a change in the environmental conditions from one place to another in the study area (for example, if one area has a higher elevation, or has more water). If one area was exposed to forest fire or a disease, adult pines might be less numerous in this area, affecting your sampling.

2.   In order to develop an effective management plan and set catch limits for a commercially important species of fishes, ecologists need a lot of information about the species. For example, what is the percentage of individuals of various ages in the population? Is there seasonal migration? What time of year do female fish spawn? (You wouldn't want to catch any during this period.) This and other life history information can be obtained from mark and recapture data.

Landfill Report

Names of Group Members: **Laboratory Section:** _____

_____   _____

_____

1.   Copy the results of your experiment into the table below. **Place an asterisk** next to the results supplied by your **partner group**.

Results of Landfill Experiment				
Item	Initial Surface Area:	Initial Surface Area:	Final Surface Area:	Final Surface Area:
	Average Value for Dry Samples	Average Value for Moist Samples	Average Value for Dry Samples	Average Value for Moist Samples
Wood chips				
Popcorn				
Paper bags				
Aluminum foil				
Styrofoam packing material				
Newspaper				
Plastic bags				
Cornstarch packing material				

2.   **(Circle one answer.)**

My hypotheses about which items would degrade were **supported/not supported**.

Explain your answer in detail, mentioning **each hypothesis and each of the eight types** of household materials.

3.   How did the presence of **moisture** affect decomposition of the **eight materials**? **Be specific**. Back up your statements by citing **facts** collected in the experiment.

4.   From the results of your experiment, what can you **conclude** about the decomposition of various trash items in a landfill?

Support your conclusions by citing **facts** from your experiments.

5.   Are **living organisms** involved in decomposition? _____

If so, what types of organisms may have been involved in the decomposition of the materials in **your** landfills?

6.   To examine the decomposition process in more detail, you put together a land-fill in a screw cap jar (similar to landfills you constructed in class). Your landfill contained leftover food from dinner, soil, and water. All the ingredients in the landfill, including **the food, the soil, the jar, the lid, and the water were thoroughly and completely sterilized!** Do you expect that the food will decompose in your landfill? _____
Explain your answer.

7.   On the basis of the results of your landfill experiments, do you predict that decomposition would take place at the same rate in the desert as it does in the rain forest? _____
Explain your answer.

Photo Credits